Lecture Notes in Computer Science 14550

Founding Editors

Gerhard Goos
Juris Hartmanis

The series Lecture Notes in Computer Science (LNCS), including its subseries Lecture Notes in Artificial Intelligence (LNAI) and Lecture Notes in Bioinformatics (LNBI), has established itself as a medium for the publication of new developments in computer science and information technology research, teaching, and education.

LNCS enjoys close cooperation with the computer science R & D community, the series counts many renowned academics among its volume editors and paper authors, and collaborates with prestigious societies. Its mission is to serve this international community by providing an invaluable service, mainly focused on the publication of conference and workshop proceedings and postproceedings. LNCS commenced publication in 1973.

Dirk Beyer · Arnd Hartmanns · Fabrice Kordon
Editors

TOOLympics Challenge 2023

Updates, Results, Successes of the
Formal-Methods Competitions

 Springer

Editors
Dirk Beyer 🅾
LMU
Munich, Germany

Arnd Hartmanns 🅾
University of Twente
Enschede, The Netherlands

Fabrice Kordon 🅾
LIP6
Paris, France

ISSN 0302-9743   ISSN 1611-3349 (electronic)
Lecture Notes in Computer Science
ISBN 978-3-031-67694-9   ISBN 978-3-031-67695-6 (eBook)
https://doi.org/10.1007/978-3-031-67695-6

Preface

TOOLympics 2023 was the third edition of a series of events to showcase competitions in the area of formal methods. It was preceded by TOOLympics 2019 in Prague as part of ETAPS, and by the "FLoC Olympic Games" in 2014. The TOOLympics 2023 event was part of the 26th European Joint Conferences on Theory and Practice of Software (ETAPS 2023), held on April 22–27 in Paris, France.

The goal of TOOLympics is to acknowledge the achievements of the various research competitions and comparative evaluations broadly related to the field of formal methods, to explain to the audience which tools from the field of formal methods they evaluate, and to understand their commonalities and differences. The developers of the participating tools typically participate in the competitions and evaluations, choosing the right parameters for the tools, or the best workflow for the approach.

A total of ten competitions joined TOOLympics in 2023 and were presented at the event: CHC-COMP (presented by Hossein Hojjat), MCC (presented by Fabrice Kordon), QComp (presented by Arnd Hartmanns), ARCH-COMP (presented by Arnd Hartmanns), RERS (presented by Falk Howar), SL-COMP (presented by Mihaela Sighireanu), SV-COMP (presented by Dirk Beyer), Test-Comp (presented by Dirk Beyer), VerifyThis (presented by Gidon Ernst), and VT-LTC (presented by Gidon Ernst). Six of the above-mentioned competitions are represented in this proceedings volume as papers: ARCH-COMP, CHC-COMP, MCC, QComp, VerifyThis, and the VerifyThis Long-Term Challenge. Each of these papers was peer-reviewed by three independent experts, who provided many insightful comments and suggestions for improvement that were used to update the contributions for their final versions collected here. The review process was single-anonymous, that is, the reviewers were not known to the authors.

We would like to thank all organizers of competitions in the field of formal methods, in particular those who presented their competition as part of TOOLympics. We would also like to thank the ETAPS 2023 organization committee for accommodating TOOLympics—especially its general chairs Fabrice Kordon and Laure Petrucci and the chair of the ETAPS 2023 executive board, Marieke Huisman—as well as the team at Springer for the flexible publication schedule.

April 2024

Dirk Beyer
Arnd Hartmanns
Fabrice Kordon

Organization

Board of Reviewers

Contents

The ARCH-COMP Friendly Verification Competition for Continuous and Hybrid Systems

Alessandro Abate[1], Matthias Althoff[7], Lei Bu[12], Gidon Ernst[2], Goran Frehse[3(✉)], Luca Geretti[11], Taylor T. Johnson[10], Claudio Menghi[8,9], Stefan Mitsch[6], Stefan Schupp[4], and Sadegh Soudjani[5]

[1] University of Oxford, Oxford, UK
Alessandro.Abate@cs.ox.ac.uk
[2] LMU Munich, Munich, Germany
gidon.ernst@lmu.de
[3] ENSTA Paris, Institut Polytechnique de Paris, Palaiseau, France
goran.frehse@ensta-paris.fr
[4] TU Wien, Vienna, Austria
stefan.schupp@tuwien.ac.at
[5] Max Planck Institute for Software Systems, Kaiserslautern, Germany
sadegh@mpi-sws.org
[6] DePaul University, Chicago, USA
smitsch@depaul.edu
[7] Technical University of Munich, Munich, Germany
althoff@tum.de
[8] University of Bergamo, Bergamo, Italy
claudio.menghi@unibg.it
[9] McMaster University, Hamilton, Canada
menghic@mcmaster.ca
[10] Vanderbilt University, Nashville, TN, USA
taylor.johnson@vanderbilt.edu
[11] University of Verona, Verona, Italy
luca.geretti@univr.it
[12] Nanjing University, Nanjing, Jiangsu, People's Republic of China
bulei@nju.edu.cn

Abstract. The workshop on Applied Verification for Continuous and Hybrid Systems (ARCH) is an annual venue for researchers and practitioners working on automated analysis and verification of hybrid systems. ARCH-COMP is a friendly competition held with the ARCH event. The competition was established in 2017 and aims to explore, document, and push forward the state of the art in the field. It evaluates and compares methods and tools for automated hybrid systems analysis and verification on predefined benchmark problems. It is supported by an active community around several categories of problems, including linear and nonlinear systems, simulation-based and analytic methods, and models from many application domains, such as automotive systems or neural networks. This paper describes the format of the competition and its

D. Beyer et al. (Eds.): TOOLympics 2024, LNCS 14550, pp. 1–37, 2025.
https://doi.org/10.1007/978-3-031-67695-6_1

organization. It documents the experiences and decisions from the current and past editions of the competition and presents reflections and lessons learned.

Website: https://cps-vo.org/group/ARCH/FriendlyCompetition

Keywords: hybrid systems · competition · tool evaluation

1   Introduction

The design of systems with continuous, real-valued quantities can be complicated when hard constraints on their behaviors must be satisfied. For example, designing a cruise controller to maintain the vehicle at target speed while avoiding collisions is complicated due to event-based phenomena in the plant or the controller. Hybrid systems are a convenient modeling formalism to describe systems containing continuous variables subject to event-based dynamics changes. A hybrid system can model a cyber-physical system requiring a software controller to interact with a physical plant to ensure hard constraints. Examples of hard constraints on hybrid systems include assuring a minimum battery charge, maximum voltage levels in an electrical grid, or adherence to security features like a dead man's switch. Formal verification is an approach to assess whether a design satisfies hard constraints based on a mathematically rigorous analysis of a formal model of the system. It has been successfully applied to software and hardware systems since the 80 s, and researchers have been developing verification techniques for continuous and hybrid systems since the 90 s.

Verifying continuous and hybrid systems is quite challenging: even reachability is undecidable in many cases [76], and computing accurate solutions can be prohibitively expensive. The absence of sharp theoretical bounds on the computational time lends particular weight to experimental evaluations. Much of the research has therefore gone into finding abstractions and heuristics with a suitable trade-off between accuracy and computational cost such that problem instances of practical interest can be handled. However, evaluating, assessing, and comparing existing techniques raises several questions:

- Which problem instances are of practical interest, and how to find them?
- How much accuracy is really needed?
- How much computation is acceptable?

Answering these questions is challenging: The answers depend on the application domain and the particular system, the specification to be verified, and the available computation hardware. They also evolve as algorithms become more sophisticated, computation hardware gets faster, and practitioners develop more demanding problems. The goal of the ARCH workshop series and the ARCH-COMP friendly competition is to provide such problem instances to researchers and maintain a forum where the advantages and drawbacks of different approaches can be assessed. This requires defining a fair way to execute tools on problem instances and identifying instances suitable for different tools. The format and organization of ARCH-COMP are intended to facilitate progress toward these objectives and will be described in more detail in the next section.

This paper is structured as follows. Section 2 provides background information on the principles of formal verification and how they apply to continuous and hybrid systems. Section 3 describes the competition format and its organization. Section 4 presents the different tracks of the competition. Section 5 describes efforts towards repeatable experiments and results. Section 6 summarizes the achievements of ARCH-COMP, our reflections, lessons learned, and outlook.

2   Verification of Continuous, Hybrid, and Stochastic Systems

We summarize the principles of formal verification and how they apply to continuous and hybrid systems. As we will see, several difficulties arise when moving from the classic setting of finite state machines to a continuous time.

2.1   The Formal Verification Approach

A formal verification requires a model of the system and a formal description of the specification. If the formal verification procedure terminates, it either

- claims that the system satisfies the specification, possibly providing a *certificate* such as an invariant or an interpolant;
- claims that the system violates the specification, often providing a *witness* such as a counterexample;
- or returns "unknown", typically because the runtime exceeds a given upper bound.

In many verification competitions such as SV-COMP [29], participants submit their software tools, and the organizers run them on a set of benchmarks unknown a priori to the participants. The tools are evaluated by considering the correctness of their verdicts, the number of problems handled in a given time frame, and the overall runtime. To decide whether the result is correct, the problem instance either has a result known by construction, has a result considered trustworthy from a reference tool, or a referee tool checks whether the certificate or witness can be validated.

Several problems arise when replicating this procedure with continuous and hybrid systems as it is the goal in ARCH-COMP. Before discussing these problems, we define what we mean by continuous and hybrid systems and their verification.

2.2   Continuous, Hybrid, and Stochastic Systems

Continuous Systems. We consider continuous systems to be systems whose state can be described by a vector $x \in \mathbb{R}^n$, i.e., with n real-valued variables. In a *discrete-time system*, the state evolves according to a function

$$x_{k+1} = f(x_k, u_k), \qquad x_0 \in \mathcal{X}_0, u_k \in \mathcal{U},$$

where k represents time, $\mathcal{X}_0 \subseteq \mathbb{R}^n$ the set of initial conditions, and $u_k \in \mathbb{R}^m$ is exogenous to the system. This is sometimes described as a nondeterministic input, but should not be confused with the notion of inputs in control systems. The input u_k is non-deterministically chosen from the compact set $\mathcal{U} \subseteq \mathbb{R}^m$. This allows one to, e.g., model bounded disturbances or to account for the difference between the dynamics function $f(\cdot)$ and the actual system. The specification is considered to be satisfied if it is valid for all admissible sequences of u_k. The input u effectively turns the equation into an inclusion.

We consider *continuous-time systems* described by ordinary differential equations (ODEs) of the form

$$\dot{x} = f(x, u), \qquad x(0) \in \mathcal{X}_0, u(t) \in \mathcal{U},$$

where $t \in \mathbb{R}^{\geq 0}$ represents time and whose solution is a *trajectory* $x : \mathbb{R}^{\geq 0} \mapsto \mathbb{R}^n$. In some cases, trajectories may only be defined for a finite time horizon. The *type* of system is named after the type of f. If f is linear in x and u, we speak of a *linear system*, etc. As above, the input u effectively turns the equation into a differential inclusion.

Hybrid Systems. We consider hybrid systems modeled by extending finite state automata with a continuous system in each discrete state. We highlight two important aspects of this extension below, see [104] for the full definition:

– Trajectories within a discrete state must satisfy at all times a so-called *staying condition* or *invariant* $\mathcal{I}$ associated with the discrete state, i.e., $x(\tau) \in \mathcal{I}$ for all time points $\tau \in [0, t]$ in the interval from 0 to time horizon t.
– Transitions from one discrete state to another are only allowed when the continuous state x satisfies an associated *guard* condition $\mathcal{G}$, i.e., $x \in \mathcal{G}$. Note that the system may nonetheless remain in the discrete state as long as its staying condition is satisfied. A transition may instantaneously update the continuous state according to its *reset function*.

The combined effect of staying and guard conditions is to introduce non-determinism in the timing of state changes. Such nondeterminism can abstract complex or partially unknown real systems with crisp and clean formal models. However, it can also make verifying such systems very difficult or even simulating them numerically. The semantics of hybrid systems are given by *executions*, which are sequences of trajectories associated with discrete states such that each trajectory satisfies the staying condition and transitions (instantaneous state changes) satisfy the guard conditions and reset functions. An alternative way of modeling hybrid systems is with *hybrid programs* and *hybrid games* that mix discrete statements, differential equations, and potentially adversarial dynamics in an imperative programming language.

In the remainder of the paper, we will not distinguish between discrete and continuous states (unless absolutely necessary) and use the symbol x to denote the state of either continuous, discrete, or hybrid systems.

Stochastic Systems. We consider stochastic systems that are affected by uncertainty with known probability distributions. The system's behavior can be captured via a state vector $x \in \mathbb{R}^n$, i.e., with n real-valued variables. In a *discrete-time stochastic system*, the state evolves according to the stochastic difference equation

$$x_{k+1} = f(x_k, u_k, w_k), \qquad x_0 \in \mathcal{X}_0, u_k \in \mathcal{U}, \quad k = 0, 1, 2, \ldots \qquad (1)$$

where k represents time, $u_k \in \mathcal{U}$ is the input, and $(w_0, w_1, w_2, \ldots)$ is a sequence of independent and identically distributed (iid) random variables representing the uncertainty affecting the evolution of the system. In a *continuous-time stochastic system*, the state evolves according to the stochastic differential equation

$$dx = f(x, u)dt + g(x, u)dW_t, \qquad x_0 \in \mathcal{X}_0, u \in \mathcal{U}, \qquad (2)$$

where time changes continuously, $u \in \mathcal{U}$ is the input, $f(x, u)$ is called the drift, $g(x, u)$ is called the diffusion, and W_t is the Brownian motion representing the stochastic uncertainty.

Equations (1), (2) represent the simplest structure of models studied in the ARCH Stochastic Category. In general, the stochastic models could have the following different features:

- The timeline could be discrete or continuous.
- The state space could be continuous or hybrid.
- The drift in the differential equation could be linear, piecewise linear, or nonlinear.
- The noise could be Brownian motion or iid. The type of control actions affecting the system could also be different. The control actions could appear as switching signals, or changing the drift of the differential equation, or a combination of these two cases.

2.3   Verification Problems

Most of the problem instances in ARCH-COMP can be characterized as *safety problems*: Given a set of initial states $\mathcal{X}_0$ and a set of *forbidden* or *unsafe* states $\mathcal{F}$, check whether the state remains at all times outside of $\mathcal{F}$. If the system is unsafe, there exists an execution such that for some state x during this execution, $x \in \mathcal{F}$. We call such an execution a *witness* of the violation.

Safety properties can express complex properties (such as bounded liveness) that can be encoded in a so-called *observer* automaton. An observer has an error state that is reachable if (and only if) the property is violated. Setting $\mathcal{F}$ as this error state, safety is equivalent to satisfying the property.

In the case of stochastic models, as those in Eqs. (1) and (2) above, the verification problem is modified into finding the probability of satisfaction of a temporal specification (such as, in the easiest instance, safety); or checking whether such likelihood meets a user-defined threshold in the unit interval.

Set Propagation, Reachability, and Model Checking. Reachability algorithms are used to compute a finite representation of a set of states that cover all reachable states; note that there are typically infinitely many reachable states in the dense state space. This may be done similarly to that by which ODEs are solved numerically, i.e., iteratively computing one-step successors. We call this process *set propagation*, and it proceeds as follows: Starting from the set of initial states, a successor set is computed that covers the next states over a given time interval or over the next discrete transition. This process is repeated iteratively to generate, in a tree-like fashion, sets that cover all reachable states. In principle, this tree can be infinite because discrete transitions enter a cycle or continuous time diverges (assuming we only cover finite time intervals in each step). However, if a set is already covered by previously used initial sets, its successors do not need to be computed again. If all sets are eventually covered, the tree can be finite even though executions over infinite time and infinitely many transitions are covered. Checking for cover by predecessors amounts to a fixed-point check and constitutes a rudimentary form of *model checking*.

For all but the most simple dynamics, it is impossible to cover the solutions of ODEs with a single computable set, even over small time intervals. Instead, the cover is approximate, and the degree of overapproximation (ratio between excess and total states covered) is difficult to quantify precisely, particularly in a hybrid system. We call these covers *approximate reach sets*.

In the case of stochastic models, as those in Eqs. (1) and (2), general verification of temporally extended specifications boils down to computing state-dependent likelihoods, namely computing the probability that trajectories, initialized anywhere in the state space, verify the given temporal specification. This is attained via dynamic programming algorithms, namely leveraging Bellman iterations.

Decision vs Reach Set Approximation. In principle, the job of a verification tool is to decide if the specification is satisfied. In continuous and hybrid systems, the shape and size of the computed approximate reach sets contain plenty of information about the system, and it is common for engineers to plot and inspect them visually. Most verification tools can produce an approximate reach set as part of the output, in addition to the satisfaction of the property.

Bounded vs Unbounded Problems. In software verification, it is common to distinguish between bounded and unbounded instances. The bound refers to the number of discrete time steps (clock ticks) considered. In a hybrid system, we must distinguish between two types of boundedness that are orthogonal and sometimes confused in the literature:

- *bounded time problems* consider executions or trajectories whose total length, measured by time (imagine a clock running in parallel to the system), do not exceed a given bound,
- *bounded transition problems* consider executions with a bounded number of discrete state changes.

In the above classification, the "bounded model checking" problems from software verification are bounded transition problems. A bounded transition problem may be unbounded in time. A bounded time problem can correspond to an unbounded number of discrete transitions. Even in simple hybrid systems, a system may take infinite discrete transitions in a bounded time interval since discrete transitions are considered instantaneous in (discrete or continuous) time.

Some confusion can arise because the continuous trajectories are only approximated up to a given upper bound in time in most reachability algorithms. This is for practical reasons: The trajectories are covered with sets, each covering a finite time interval. Since only a finite number of sets can be computed, this limits the time horizon to the sum of the finite time intervals. However, this limitation can be sidestepped entirely if we admit an unbounded number of transitions. The trick is simple: add a clock variable to the system and impose the staying condition to include an arbitrary upper bound $T > 0$ on the clock, as well as self-loop transitions to all discrete states (self-loops do not modify the discrete state); let each self-loop reset the clock to zero. The augmented system thus obtained has exactly the same behavior as the original system, except that the clock ensures that the system does not remain in a discrete state longer than T time before taking a transition. Therefore, it is sufficient to cover continuous-time trajectories up to time T.

To summarize, most tools cover only bounded continuous time intervals between discrete transitions. Many tools consider only bounded transitions, i.e., they do the equivalent of "bounded model checking". All tools that can handle unbounded numbers of transitions, e.g., through fixed-point checking, can also handle problems that are unbounded in time.

Theorem Proving. Unlike numerical reachability analysis tools, hybrid systems theorem provers use logical reasoning to analyze fully symbolic and parametric models with unbounded initial sets and for unbounded time. Depending on the expressiveness of the underlying logics, hybrid systems theorem provers can analyze safety properties, liveness properties, stability properties, and game properties. Theorem provers attempt to construct proofs from axioms of programs, differential equations, and arithmetic. Such proofs can typically be conducted interactively, steered with tactics, or attempted fully automatically.

For differential equation analysis, instead of computing reachable sets numerically, theorem provers often use invariant techniques, such as Lyapunov functions and Barrier certificates, to prove that dynamics stay inside certain safe regions. For liveness analysis, progress properties are used. Stability analysis can be expressed as a combination of safety and liveness [168]. Highly trustworthy theorem provers separate searching for such invariant properties from certifying them from axioms, which enables the use of untrusted numerical procedures during search; some theorem provers use invariant search directly as part of their trusted code base or defer this task to the user. In addition to establishing correctness properties about models, auxiliary development tasks, such as stepwise refinement [133], runtime monitoring [146], compilation of models to executable code [35, 92, 167], and derivation of artifacts for machine learning [90, 156], are expressible in logic and supported by some theorem provers.

2.4   Synthesis Problems

Many of the previously mentioned systems exhibit *non-determinism* in the form of uncertainty and/or control actions. For the analysis of stochastic systems, it is relevant to resolve this non-determinism to analyze the system's properties.

Different approaches, often referred to as *schedulers*, *policies*, or *controllers*, exist to resolve non-determinism that affects the behavior of the resulting system. For instance, stochastic schedulers resolve non-determinism according to a given distribution. State-dependent schedulers, on the other hand, resolve non-determinism based on the current system state. A richer class is given by history-dependent schedulers, which incorporate not only the current state but also past evolution to resolve non-deterministic choices.

In this setup, the synthesis problem demands a scheduler such that the probability of satisfying a given specification is maximized, an expected total cost is minimized, or an expected total reward is maximized.

Note that the synthesis problem is much more complex than the verification problem since we search for optimal schedulers on a possibly uncountable functional space. Therefore, formal and sound solutions are developed to synthesize schedulers with correctness guarantees. The main challenges to consider include: (a) the given specification to be checked on the model could have a finite or infinite time horizon (characterizing the infinite-horizon behavior of the system is more complicated than the finite-horizon behavior); (b) The required specification could go beyond safety and reachability, which induces augmented hybrid models with both continuous and discrete state components; (c) Resolving the probabilistic non-determinism is difficult on the considered benchmarks since the system could be influenced by Brownian motions and Poisson processes; and (d) The evolution of the (augmented) system could be derived from both deterministic and probabilistic equations.

In the case of stochastic models, as those in Eqs. (1) and (2), general synthesis goals reduce to computing optimal policies using Bellman iterations - this is a slight generalization of the approaches and algorithms for verification described above.

2.5   Problem Instances

We distinguish three types of problem instances, each serving a different purpose:

- *Toy problems* mainly serve educational purposes. However, some, such as the bouncing ball, showcase fundamental properties: How reach sets are approximated, how the approximation error and the number of sets increase with each transition, whether there's a fixed-point check, etc. So, to experts, certain toy problems can be quite representative and useful for developing new software tools.

- *Academic benchmarks* serve to illustrate a particular aspect. For instance, a parametric benchmark with varying numbers of discrete states can be useful to compare different tools on this aspect.
- *Industrial benchmarks* ideally originate from actual design problems in application domains. They constitute a Litmus test of whether tools can handle problems of practical interest. However, they are hard to come by and can be biased towards certain classes of problems. They also tend to be either excessively hard or too easy for effective comparisons.

In ARCH-COMP, we use all three types of problem instances to assess tools on a wide range of criteria, while also evaluating their capability to solve problems of practical or industrial interest.

2.6   Inherent Challenges in Evaluating Results

When evaluating verification tools for continuous and hybrid systems in the spirit of the formal verification approach outlined in Sect. 2.1 there are fundamental and practical problems.

Undecidable Certificates. The role of certificates is to provide a way to check that a system satisfies a property. In continuous and hybrid systems, even rudimentary problems such as verifying an invariant in a single discrete state are generally undecidable. Thus, checking the validity of a certificate can be as difficult as checking the property itself. So far, we have not used certificates in ARCH-COMP. However, we use visualizations of reach set approximations for sanity checks, which can be considered a first step in this direction.

Missing Witnesses. A witness is a collection of data, e.g., an execution, that allows one to check that a property is violated by a system. For continuous and hybrid systems, this is problematic in a fundamental way. For all but the simplest classes, trajectories can be described at best by transcendental functions, such as exponential functions of time. The sequence of states that describe a witness would almost surely involve irrational numbers. More practically speaking, a witness produced by a verification can be refuted by a referee tool, but it rarely can be confirmed as an actual execution of the system. Therefore, finding a cover guaranteed to contain at least one solution can be helpful. However, no tool currently outputs such a witness cover.

Differing Semantics. Various modeling and specification formalisms have been proposed for continuous and hybrid systems. A survey from 2005 mentions no less than five fundamentally different semantics for hybrid automata alone [87], and other formalisms for hybrid systems abound [27]. While some differences are more technical, others directly affect whether a problem instance can be effectively characterized and the verification outcome. We have sometimes included different variations, such as continuous-time and discrete-time instances, for inclusivity. Similarly, the stochastic model category has seen the presence of very diverse and semantically rich modeling formalisms and corresponding benchmarks that are not always easy to adapt to other models.

Missing Exchange Format. Because of the multitude of different classes of models and properties, there is no readily applicable way to specify problem instances in a uniform manner. By way of a compromise, some tools have adopted a common syntax based on the hybrid automaton formalism. Efforts to define interchange formats have not seen widespread success [26]. De facto, each tool has its way of specifying models and properties. This is particularly true in the instance of stochastic models.

Hand-Crafted Hyperparameters. Most verification algorithms involve many hyperparameters, such as the time step used for reach set approximation, parameters to define set approximations (orders of zonotopes or Taylor models), and other ways to abstract from or simplify representations of sets of behaviors. Some attempts at automating the choice of hyperparameters have been made [21, 175–179], but in general, the proper way to use many tools remains expert knowledge that is not easily obtained or widely disseminated.

3   Competition Format and Organization

The main objective of ARCH-COMP is to evaluate and compare different approaches across a set of benchmark problems. While performance measures such as time and memory consumption are evaluated, we consider them secondary to the test of whether participants can solve interesting problems – call it the Olympic spirit if you like.

The competition is accompanied by a workshop, which provides a means to propose and present benchmark problems and showcase tools in detail. The workshop maintains a curated benchmark repository, where models and specifications are archived and can be updated where necessary. The accompanying discussions occur in a public electronic forum associated with the workshop.

3.1   A Friendly Format

As we described in Sect. 2.6, we are confronted with many problem classes, diverse solution methods, and a lack of unified interfaces and hyperparameters. Consequently, we opt for the format of a *friendly competition*: Participants create and select problem instances and can tune hyperparameters for each instance before submitting their solution instances in a repeatability package. This ensures that each tool is showcased under its most suitable configuration, which is not necessarily the case when running with a default configuration or when hyperparameters are chosen by the organizers instead.

We have two mechanisms to assure that solution instances are indeed true solutions and not the result of cheating (which we never had) or the lucky result of modeling or specification errors (of which we had several):

– Solution instances are accessible to other participants during the competition and publicly archived afterward. An impressive performance is, therefore, likely to provoke scrutiny by experts. This incentivizes participants to submit solutions that can withstand such scrutiny over time.

– Problem instances can be designed to bracket the computed solutions. Take a problem instance where the model satisfies the specification. A *bracket instance* would be a slight modification of either the model or the specification such that the specification is now violated. If a tool gives correct answers for both bracket instances *with the same set of hyperparameters*, the tool's accuracy lies within the solution space left by the bracket instances. For some classes of problems (like convex invariants), this could be formalized so that correctly solving a set of bracket instances formally guarantees a given level of accuracy of the tool. So far, we have not felt the need to take it to such a formal level.

3.2   Organization and Schedule

The competition is divided into groups that address different problems and will be described in more detail in Sect. 4. Each group is managed by a group leader, who organizes discussion rounds and the joint writing of a final report for each group. All decisions are taken by consensus. This decentralized organization allows each group to develop rules and criteria suitable for their problems.

Each year, the competition follows the following schedule:

1. Following a public call for participation, participants register with a group.
2. The group meets to select problem instances and propose new ones. Groups are encouraged to include benchmarks from the ARCH proceedings.
3. Participants submit preliminary results discussed in a second group meeting. Difficulties and misunderstandings can be resolved in this phase, and surprise performances can be discussed. If necessary, problem instances are refined or clarified.
4. Participants submit their final solution instances in a repeatability package. The evaluation chair runs all packages, and the resulting performance logs are returned to participants. The packages are publicly archived.
5. Each group writes a final report, which includes descriptions of problem instances, performance results, and a discussion of those results. The reports are published in the ARCH proceedings.
6. The competition closes with a presentation by each group leader, given at the ARCH workshop. The audience votes for a winning tool in each group and casts a final vote for the overall most impressive result.

3.3   Artifacts and Results

Problem instances are described in natural language in the ARCH workshop group reports or benchmark papers. Where possible, formal models for each instance are also provided in an online repository in formats recognized by the community [9].

Each participating tool deposits its artifacts for solving the problem instances in the repository (program code or executables, scripts, configuration files, etc.) and a script that runs each instance and provides the result in a given format.

Typically, the result is whether the specification is satisfied, not satisfied, or the instance cannot be decided. This script is used in the repeatability evaluation, which will be described in Sect. 5, and to measure the runtimes and memory consumption.

Depending on the group and the problem instances, tools also produce additional output, such as plots of reachable states. We have found this approach helpful to gain further insight, e.g., when one tool unexpectedly outperforms another, for quick and intuitive estimation of the precision of the results, and as a sanity check for surprisingly good or bad performance. Where useful, such plots are included in the group reports.

4   Thematic Groups of the Competition

The competition is organized by groups (tracks), which operate and report independently since they address different problems and solution methods. Each group tackles a different class of models (e.g., linear vs. nonlinear), methods (reach set approximation vs. theorem proving), and objectives (verifying safety vs. falsification). This section provides a summary related to each group of the competition. For each group, we report its (i) goal, i.e., the problem addressed by the track, (ii) benchmarks, i.e., the benchmark problems considered in the competition; (iii) participants, i.e., the tools participating in the track; and (iv) outcome, i.e., a description of the outcomes produced by each track across the years. Table 1 reports the tools participating in each group for each year of the competition. Table 2 reports the number of benchmarks considered for each group and year. We do not report illustrative graphics for the results of each group since the goal of this work is to provide general reflections about the competition and not to discuss the results in detail. The reader can refer to the corresponding group reports for this analysis.

4.1   Piecewise Constant Dynamics

Goal. In ARCH-COMP, we have a track PCDB for *continuous and hybrid systems with piecewise constant dynamics* (HPCD) and *bounded model checking* (BMC) of HPCD systems. The PCDB category concerns hybrid systems where in each location (mode, piece of the hybrid state space), the dynamics are given by a differential inclusion of the form $\dot{x}(t) \in \mathcal{U}$, where $\mathcal{U}$ is a convex subset of $\mathbb{R}^n$. Specifically, the BMC task concerns the bounded model checking of HPCD systems where the bound is described as the depth of the discrete jump of the system. The verification specifications used in PCDB track focus on the (bounded) reachability verification, e.g., whether two processes can be in the critical section at the same time, or whether a vehicle can reach a specific dangerous position, and so on. As the main techniques, implementations used by different checkers vary from each other, the goal of the PCDB track is to present the landscape of existing solutions in a breadth and showcase the current state of the art.

Table 1. Tools participating in each group for each year of the competition

Category	Participating Tools						
	2017	2018	2019	2020	2021	2022	2023
Piecewise Constant Dynamics	BACH [43,44], Lyse [33], PHAVer/SX [82], VeriSiMPL [6,7].	BACH [43,44], Lyse [33], PHAVer/SX [82], PHAVerLite [25], VeriSiMPL [6,7].	BACH [43,44], HyCOMP [55], Lyse [33], PHAVer/SX [82], PHAVerLite [25], VeriSiMPL [6,7].	BACH [43,44], PHAVer/SX [82], PHAVerLite [25], ROPICAL [149], XSpeed [158].	N/A	BACH [43,44], PHAVer/SX [82], PHAVerLite [25], SAT-Reach [158], XSpeed [158].	N/A
Linear Dynamics	Axelerator [49] CORA [10] Flow* [51] HyDRA [161] Hylaa [22] SpaceEx [83] XSpeed [158]	CORA [10] CORA/SX [14] C2E2 [74] Flow* [51] HyDRA [161] Hylaa [22] JuliaReach [31] SpaceEx [83] XSpeed [158]	CORA [10] CORA/SX [14] HyDRA [161] Hylaa [22] JuliaReach [31] SpaceEx [83] XSpeed [158]	CORA [10] C2E2 [74] HyDRA [161] Hylaa [22] Hylaa-Continuous [23] JuliaReach [31] SpaceEx [83] XSpeed [158]	CORA [10] HyDRA [161] JuliaReach [31] SpaceEx [83]	CORA [10] JuliaReach [31]	CORA [10] JuliaReach [31] Verse [132].
Nonlinear Dynamics	CORA [10] Flow* [51] Isabelle /HOL [109]	CORA [10] CORA/SX [14] C2E2 [74] Flow* [51] Isabelle /HOL [109] SymReach [72]	Ariadne [40] CORA [10] DynIbex [160] Flow* [51] Isabelle /HOL [109] JuliaReach [31]	Ariadne [40] CORA [10] DynIbex [160] Flow* [51] Isabelle /HOL [109] JuliaReach [31]	Ariadne [40] CORA [10] DynIbex [160] JuliaReach [31] Kaa [125]	Ariadne [40] CORA [10] DynIbex [160] JuliaReach [31] KeYmaera X [89]	Ariadne [40] CORA [10] DynIbex [160] JuliaReach [31] KeYmaera X [89] Verse [132] .
AINNCS	N/A	N/A	NNV [136,170], Sherlock [63–65], Verisig [112,113],	NNV [136,170], OVERT [164], ReachNN* [75,106], VenMAS [8].	JuliaReach [31], NNV [136,170], Verisig [112,113],	CORA [10,126], JuliaReach [31], NNV [136,170], POLAR [105].	CORA [10,126], JuliaReach [31], NNV [136,170].
Stochastic Models	N/A	Barrier [114], FAUST² [166], FIRM − GDTL [131], Modest [103], SDCPN&IPS [71], SReachTools [172].	FAUST² [166], HYPEG [153], LyapMMC [159], Modest [103], SDCPN&IPS [71], SReachTools [172], StocHy [50], SySCoRe [102].	AMYTISS [129], FAUST² [166], hpnmg [108], Mascot-SDS [140], Modest [103], ProbReach [163], SDCPN&IPS [71, 138], SReachTools [172], StocHy [50].	AMYTISS [129], FIGARO [36,37], hpnmg [108], HYPEG [153], Mascot-SDS [139, 140], Modest [103], ProbReach [163], PyCATSHOO [53], SDCPN&IPS [71, 138], SReachTools [172], StocHy [50], SySCoRe [101].	FIGARO [36,37], HYPEG [153], Modest [103], RealySt [58], PyCATSHOO [53], SDCPN&IPS [71, 138], SySCoRe [101].	HYPEG [153], ProbReach [163], RealySt [58], SySCoRe [171].
Falsification	S-TaLiRo [19].	S-TaLiRo [19], FalStar [70].	Breach [61], FalStar [70], falsify [181], S-TaLiRo [19].	ARIsTEO [142], Breach [61], falsify [181], FalStar [70], S-TaLiRo [19], zlscheck [183].	ARIsTEO [142], Breach [61], FalCAuN [173], falsify [181], FalStar [70], ForeSee [182], S-TaLiRo [19], Ψ-TaLiRo [169].	ARIsTEO [142], FalCAuN [173], falsify [181], FalStar [70], ForeSee [182], S-TaLiRo [19], Ψ-TaLiRo [169].	ARIsTEO [142], ATheNA [78], FalCAuN [173], ForeSee [182], NNFal [151], STGEM [152], S-TaLiRo [19], Ψ-TaLiRo [169].
Hybrid Systems Theorem Proving	N/A	KeYmaera X [89] KeYmaera 3 [155], HHL Prover [174]	KeYmaera X [89], KeYmaera 3 [155], HHL Prover [174]	KeYmaera X [89], KeYmaera 3 [155], HHL Prover [174], HybridVCs [150]	KeYmaera X [89], HHL Prover [174]	KeYmaera X [89], HHL Prover [174], HHLPy [162], IsaVODEs [79]	KeYmaera X [89], HHL Prover [174], HHLPy [162], IsaVODEs [79]

Benchmarks. The benchmark collection has evolved continuously with each edition of the competiton [41, 42, 45–47, 84–86]. Since HPCD and BMC are two parallel tracks at the beginning, and merged to PCDB in the edition of 2020, the benchmarks contain both unbounded and bounded specifications for each cases. Most of the benchmarks are extendable, so that the models can be concretized with different values of parameters to increase the difficulty of the problem in the aspects of number of continuous variables, number of discrete locations, and number of components in the system. Thus, the benchmarks can be used in the evaluation of the efficiency, scalability of different tools.

Table 2. Number of benchmark models considered for each group and year

Category	Number of benchmark Models						
	2017	2018	2019	2020	2021	2022	2023
Piecewise Constant Dynamics	5	5	5	6	N/A	6	N/A
Linear Dynamics	3	6	6	8	9	9	8
Nonlinear Dynamics	3	4	4	6	5	6	6
AINNCS	N/A	N/A	5	7	7	10	10
Stochastic Models	N/A	5	4	7	10	6	2
Falsification	1	1	6	7	7	6	7
Hybrid Systems Theorem Proving	N/A	139	169	214	214	220	221

Participants. Throughout the years, different checkers have joined the competition, while not everyone of them participated every year. These tools are using different techniques, e.g. SMT encoding and solving based, polyherdal based geometric computation, support function based verification, and so on. Past participant tools in this category in alphabetical order are BACH [43,44], HyCOMP [55], HyDra [161], Lyse [33], PHAVer/SX [82], PHAVerLite [25], SAT-Reach [158], SpaceEx [83], TROPICAL [149], VeriSiMPL [6,7] , XSpeed [158].

Outcome. In the evaluation reports, we can see the size of individual automata that can be solved has increased significantly. It will be an important topic to evaluate whether existing methods/tools can handle large compositional system efficiently. Besides of the evaluation results, the PCDB track has achieved a stable benchmark set in a well recognized format of models.

4.2 Continuous and Hybrid Systems with Linear Dynamics

Goal. While there exists an analytical solution for piecewise constant dynamics, the solution of linear systems can be computed exactly in some specific cases only, e.g., when all eigenvalues are real or imaginary [128]. However, linear dynamics can be considered the most straightforward dynamics besides piecewise constant dynamics because the superposition principle can be used, the homogeneous solution can be computed analytically using the matrix exponential, and the convexity of reachable sets of points is preserved in time. This makes it possible to use convex set representations and compute the reachable set without the wrapping effect [99]. Because the verification of purely linear systems is already fairly well understood, the goal of this track is to assess how to scale the computation using order reduction methods with formal error bounds and decomposition techniques [11,23,32]. The verification of hybrid systems with linear dynamics is much less understood. While novel ideas have already been proposed to solve the problem of precise and scalable guard intersection [16,20,98], benchmarking different concepts for guard intersections is a primary goal. Finally, a further goal is to evaluate fully automatic verification processes as proposed, e.g., in [175,177].

Benchmarks. Over the years, we have added more challenging benchmarks while some easy benchmarks have been removed [12–15, 17, 18]. In particular, we have added high-dimensional problems with up to one million state variables and removed low-dimensional ones, such as the building benchmark with 48 state variables. In general, the number of continuous state variables for hybrid systems is much lower than for purely continuous systems due to the difficulty of computing precise and scalable guard intersections. In the 2023 edition, three benchmarks were purely continuous, while five were hybrid. The hybrid benchmarks cover cases where guards trigger the discrete transitions and where the discrete transition changes can happen arbitrarily.

Participants. So far, eleven tools have participated in this category. All tools essentially use some form of reachability analysis. Past participating tools in this category in alphabetical order are Axelerator [49], CORA [10], C2E2 [62], Flow* [51], HyDRA [161], Hylaa [22], Hylaa-Continuous [23], JuliaRech [31], SpaceEx [83], Verse [132], and XSpeed [158].

Outcome. By comparing the results in the yearly competitions, each tool could improve more than without the gained insights. The number of state variables that can be considered today is several orders of magnitude larger than what we could verify in the competition's first edition (2017). Some verification results are now automatically obtained, which was realized for the first time in 2023. Besides these scientific achievements, several achievements regarding the reproducibility of results have been pioneered in this category. This includes using the same modeling format for all benchmarks (SpaceEx model definition), Docker files for executing all results and fully automatically evaluating Docker files on the same server to better compare computation times.

4.3   Nonlinear Dynamics

Goal. As opposed to piecewise constant and linear systems, nonlinear differential systems do not have an analytical solution and the superposition principle is not applicable. As a result, in the numerical case, an (over-)approximated solution is typically computed. Given the nonconvexity of sets in general, set representation is particular important. Specifically, how to provide a finite polynomial approximation that introduces a small remainder for rigorous analysis. Currently, Taylor polynomials [28] are the most commonly used representation for the enclosure of nonlinear sets, although there has been some work on Bernstein polynomials in recent years [57]. Given the sensitivity of those approximations to the problem at hand, lately some work has been done in automating numerical reachability parameters [93, 179].

Since the majority of the tools in this category fall into the numerical approach, the main goal is to show improvements in the ability to represent the finite-time reachable set as tightly as possible within reasonable computation times. While most of the problems with nonlinearity are associated to continuous evolution, hybrid evolution also introduces its own challenges. Nonlinear

guards and their intersections are not trivially understood, with respect to varying concavity and the difficulty to handle tangential crossings. Consequently,
a secondary goal has been to capture those critical cases and evaluate how to
effectively address them.

Benchmarks. The benchmarks suite has steadily evolved over the years [52, 94–
97, 110, 111]. The general policy for choosing benchmarks is to allow as many
tools as possible to return an answer to the corresponding verification problem.
Given the mixed numerical and symbolic approaches used, this objective can be
particularly daunting. As a result, some benchmarks have been adapted across
the years to become more/less challenging based on average progress from the
existing participants or the presence of new participants. To explicitly manage
this versioning, all benchmarks are currently identified by a four letter contrac-
tion of the name followed by two digits representing the most recent year they
were updated.

Recently, there has been a greater effort in two specific directions: addressing
inherent issues with numerical stability (on the continuous side) and assessing
the quality of sets after transitions (on the hybrid side). Since 2020, out of
the average six benchmarks, two are specifically hybrid to deal with transverse
crossings (LOVO21) and large sets crossings (SPRE22), respectively.

Participants. Throughout the years, 12 different tools joined the competition.
Most tools had a purely numerical approach, with a minority using symbolic
approaches. Typically there have been 6 participants each year. Past partici-
pating tools in this category in alphabetical order are Ariadne [40], C2E2 [62],
CORA [10], CORA/SX [14], DynIbex [160], Flow* [51], Isabelle/HOL [109],
JuliaReach [31], Kaa [125], KeYmaera X [89], SymReach [72] and Verse [132].

Outcome. The interaction between research groups that was solicited from the
competition (during it and outside of it) produced improvements in many tools
across the years. While there has been no particular focus on increasing the
number of variables handled, combinations of dynamics and large initial sets
originally not addressable progressively became the standard for verification.
There is still a significant amount of work to be done on numerical tools to support some categories of transitions necessary to analyze hybrid systems. On the
other side, some symbolic tools were not originally designed to work with hybrid
dynamics either, requiring extra effort to support the existing benchmarks.

4.4 Artificial Intelligence and Neural Network Control Systems (AINNCS)

Goal. Autonomous systems increasingly incorporate artificial intelligence (AI)
and machine learning components, such as neural networks (NNs), for various
sensing/measurement and control tasks ranging from perception, sensor fusion,
planning, and feedback control. Control theory, particularly in the intelligent

control area, investigated significantly the usage of neural networks as feedback controllers. This category primarily considers such neural network control systems (NNCS), where a feedback control policy is designed and implemented as a neural network providing input to some plant, the latter of which is modeled as differential equations or hybrid systems. Specifications considered have been safety properties (invariants) and some limited reachability properties. The category was first held in 2019 [137] and has been held annually since then [123, 124, 134, 135], with varying participants over the years.

Benchmarks. The benchmarks have varied over the years of the competition. Most benchmarks consist of a continuous-time and continuous-state plant modeled using linear or nonlinear ODEs, with properties—mostly safety—defined as predicates over the state-space of the plant model. Most of the plant models have been of low dimensionality, with around two to ten state variables. Typically, the neural network controller generates inputs for the plant model periodically, in a sample-and-hold fashion, so the control input produced by the neural network is applied to the plant for the entire control period. Most controllers have consisted of fairly small (orders of tens to hundreds) of neurons, mostly with ReLU activation functions. A few benchmarks have varied from this typical structure, with some in discrete time.

Participants. The participating tools have varied over the years of the AINNCS category. The tools that have participated in the AINNCS category in any year since 2019 are (in alphabetic order): COntinuous Reachability Analyzer (CORA) [10, 126], JuliaReach [31], NNV (Neural Network Verification Tool) [136, 170], Sherlock [63–65], POLAR [105], OVERT [164], ReachNN* [75, 106], VenMAS [8], and Verisig [112, 113]. One tool has participated in every iteration (NNV), and several tools have participated in two or more iterations (CORA, JuliaReach, POLAR/ReachNN*, and Verisig).

Outcomes. The AINNCS category has led to several outcomes in the verification of systems controlled by neural networks. The first major contribution is the support of this new research area and community for verification of hybrid and continuous systems that use AI components, where now nine verification tools have participated over the years. The next significant contribution is the curation of a set of around a dozen challenging benchmarks now, that have been used within the community in the development of new methods, for comparisons and motivation of new challenges. Within this has been standardization effort for the representation of the neural network controllers, specifically in the ONNX format, along with representation of the plant models in interchange formats. Another outcome is the identification of significant challenges in this space, for example, the scalability in both the sizes of the neural network controllers and the plant dimensionality currently are fairly low, and motivate the development of further scalable verification methods and effective abstractions that are not overly conservative. There are many such challenges to address for the AINNCS category in future iterations of the competition that we hope to continue in the next iterations of ARCH-COMP.

4.5   Stochastic Models

Goal. Approaches and tools in the ARCH stochastic category aim to verify and synthesize systems that combine discrete, continuous, and stochastic behavior. In this context, *verification* tries to answer questions about the probability of reachability and other specifications. An example of such specifications is to check whether "the probability to reach a subset of the state space A where variable $x \geq 3$ holds is larger than 0.8." Such specifications are encoded appropriately in formal specification languages. Furthermore, part of this category is dedicated to solving *synthesis* problems, i.e., finding a suitable resolution of existing non-determinism such that a given specification (as before) is satisfied. Apart from evaluating tools, each year the participants of the category decide on a goal that serves the community such as categorizing and classifying benchmarks or, more recently, the development of a set of toy examples in different modeling formalisms that each participating tool can solve.

Benchmarks. The set of benchmarks as described in previous reports [1–5] incorporates eleven systems from different communities, which have been collected by the participants over the years. Several benchmarks are also known from other communities but have been extended by including stochastic behaviors. Others come from an industrial application, e.g., from energy systems, robotics, or healthcare. As the approaches and goals of the participating tools vary, the collection of benchmarks exhibits a diverse set of challenges such that there is currently no tool that can solve all benchmarks.

Participants. Up to this day, 15 tools have participated in this category. While not every approach participated every year, on average five tools have taken part in the event. Furthermore, we have hosted various experts from the field, taking part in the bi-weekly meetings and contributing to discussions. Tools that participate are based on a diverse set of theoretical approaches, such as abstraction-based methods, simulation relations, coupled stochastic relations, statistical model checking, rare event simulation, kernel-embedding methods, and approaches based on reachability computation (stochastic extension of classic reachability analysis for hybrid system). We refer the reader to the paper [130] for a survey on the theoretical developments of these approaches on formal verification and synthesis of stochastic systems. Past participating tools in the ARCH stochastic category in alphabetical order are AMYTISS [129], FAUST^2 [166], Figaro [38], hpnmg [108], HYPEG [153], Mascot-SDS [139], modes [48], ProbReach [163], prohver [80], PyCATSHOO [54], RealySt [58], SDCPN&IPS [30], SReachTools [172], StocHy [50], and SySCoRe [171].

Outcome. Over the years, apart from benchmark evaluation, the ARCH stochastic category has fostered a lively exchange on community-relevant topics that has resulted in several initiatives besides the main goal of the competition. One central aspect in this regard has been the discussion of different modeling paradigms for stochastic systems, how to represent a given system within these modeling paradigms, and whether an interchange/exchange format could be developed, as done earlier for other models. This overall goal, however, is arguably harder than in other categories, in view of the semantical richness and diversity of stochastic (and additionally hybrid) modeling frameworks.

4.6   Falsification

Goals. It targets the black-/greybox analysis of executable models considering requirements expressed in temporal logic with time bounds, encoded in metric temporal logic (MTL [127]) or signal temporal logic (STL [141]) over a finite time horizon. The participants need to find initial conditions and time-varying inputs subject to certain constraints that steer the system into a violation of the respective requirement. The goal is to compare how quickly tools find witness signals for such violations, and moreover to compare the statistical variability in these results, since many approaches are stochastic.

Benchmarks. As described in past reports [59,60,66–69], the benchmark set has been growing continuously and encompasses seven system models, including some well-known models in the automotive domain that have been widely used in the literature [115]. Each comes with a number of requirements, comprised of a definition of the search space of input signals, as well as the temporal logic formula to be falsified. The models encompass a variety of difficulties, such as nonlinear or discontinuous behavior, and large search spaces.

Participants. The participants typically rely on simulation-based approaches and employ quantitative metrics [73,81] to measure how close a given input is to violating a requirement ("robustness semantics"). Research in this area has produced a variety of techniques, mature tools, and practical applications; these are described in overview survey articles [24,56]. More recently, approaches based on system identification have participated, which first learn a surrogate model of the behavior, over which the falsification problem is easier to solve. Past participants of the competition include ARIsTEO [142], ATheNA [78], Breach [61], FalCAuN [173], ForeSee [182], NNFal [151], STGEM [152], falsify [181], FalStar [70], S-TaLiRo [19], Ψ-TaLiRo [169], and zlscheck (based on Zélus [39]).

Outcome. The falsification category has achieved a stable benchmark set that is gradually becoming a standard in the falsification literature. Moreover, recently a validation step [66] has been introduced to detect any discrepancies between tools, simulation environments, model versions, and experimental setup; all major such issues have been found and fixed. This increases trust in the empirical comparison, which is available now for a wide range of tools as a reference.

4.7   Hybrid Systems Theorem Proving

Scope and Goals. The characteristic feature of the hybrid systems theorem proving category is its emphasis of programming languages as structuring principles for hybrid systems. The unambiguity and precision of program language semantics paves the way for mathematical rigor of logical reasoning principles. Typical approaches in this category perform deduction based on a program logic for hybrid systems and hybrid games, such as differential dynamic logic (dL [154]) or Hybrid Hoare Calculus [100]. Unlike other categories in the competition, hybrid systems theorem proving uses *fully symbolic, non-deterministic, parametric models*, and focuses on *infinite-time* verification from *unbounded starting states*. As a result, the proofs in this category typically identify fundamental design characteristics of the analyzed models, such as loop invariants and invariant properties of differential equations. The examples vary in scale from basic hybrid programs to industrial case studies, such as verification of collision avoidance in autonomous ground vehicles. The correctness specifications in this category express necessity (safety, programs only reach safe states), eventuality (liveness, programs eventually reach goal states), and winning strategies (games, one of the competing players wins). Future extension to stability [168] is planned.

Benchmarks. The competition benchmark set includes common design shapes at a small scale to test theorem proving base functionality, nonlinear and parametric continuous models to assess the continuous reasoning capabilities, hybrid games, and full-scale hybrid systems case studies, each in three modes:

– fully automatic verification without any additional input beyond the original hybrid system and its safety specification (in particular, without any proof scripts or other parametrization of the proof procedures);
– semi-automatic verification from design insights that are annotated to the original problem specification, allowing users to communicate specific advice about the system such as loop invariants;
– proof checking from proof scripts, which perform a significant part of the verification or provide problem-specific proof tactics.

Recent editions of the benchmarks in this category [144,147,148] use select examples to make a more detailed side-by-side comparison of modeling features and verification approaches.

Participants. Benchmark examples in this category are written in differential dynamic logic [154] which has axioms and an unambiguous semantics available in Isabelle/HOL and Coq [34], and in KeYmaera 3 [155] and KeYmaera X [89]. If known, the benchmark examples come with proof hints for semi-automatic verification and proof scripts for proof checking. A tutorial on the modeling principles in differential dynamic logic can be found in [157], whereas details on the ASCII syntax are in [145]. From this common format, participating tools translate benchmark examples into their own input syntax. Participants in this category included KeYmaera X [89] with Pegasus invariant generator [165], Bellerophon tactic script language [88], implicit definitions of functions [91], and implicit and explicit proof management [143]; IsaVODEs [79,107]; HHL Prover [174], and HHLPy [162].

Outcome. One of the benefits of hybrid systems theorem proving is its transparency in terms of human-inspectable proofs or disproofs that justify why a hybrid systems model does or does not satisfy the desired properties. Many significant theorem proving results (in general, not only for hybrid systems) were obtained with manual guidance to find such proofs, which highlights another benefit: even if fully automated tools may get stuck, theorem proving with manual guidance can still make progress. For benchmarking purposes, however, this poses a challenge when it comes to comparing the reasoning performance of hybrid systems theorem provers.

For full performance comparison transparency, the hybrid systems theorem proving category introduced the automated mode, in which tools are required to find proofs in their default configuration automatically without hints from a user. In automated mode, the number of solved examples and their duration reflect the capabilities that tools provide to novice users or users that focus on modeling, but not on proving. A challenge related to proof automation is proof portability: since tools generally integrate a mix of proof search techniques whose use is typically balanced with timeouts, the compute power of a machine searching and checking a proof may influence whether or not a proof can be found (for example, a proof search heuristic may succeed within its default timeout on one machine, but abort with a timeout on a lesser machine).

Naturally, even as automation made significant advances over the course of the competition, the number of proofs found fully automated remained lower than in interactive or hints mode, since better automation lets users become more efficient in doing manual proofs. The number of solved examples and the tactic script lengths in hints and interactive mode are indicative of the state-of-the-art performance and the user effort necessary to achieve such results.

In summary, theorem proving is a complementary technique to reachability analysis that can find concise, easily checkable, and human-interpretable correctness proofs with significantly lower computation time [96]. Progress in case studies of significant size is achievable with human guidance. Recent competition instances additionally introduced a qualitative side-by-side comparison of modeling features, semantic similarities and differences, and proof examples to provide better intuition about the mechanics of modeling and conducting proofs in different theorem proving systems.

5   Repeatability

Repeatability aims to enable the replication of results and experiments on the same (or another platform) and provides additional evidence of the validity of the results. The submission and collection of the benchmarks, execution scripts, installation scripts, and corresponding instructions may provide the capability for future researchers to build upon and reuse these computational results for other purposes, for instance, in comparisons in research papers.

The competition makes significant efforts to ensure the repeatability of the experiments. Each iteration of the competition has had a repeatability evaluation process to validate the results of the participants in the different categories, the specific details of which may be found in each iteration's repeatability evaluation report [116–122]. The general repeatability process and its goals are similar to artifact and repeatability evaluations done in some computer science conferences over the past decade or so (see e.g. [180]), to enhance trust and validity of computational results and make available computational artifacts that support claims and conclusions made through research papers and reports. The process has evolved over the years, initially beginning with a significant manual effort, to near full automation in the most recent iterations of the competition.

Generally, the evaluation has been led primarily by the evaluation chair (Taylor Johnson), who installed and reran the tools on the benchmarks across all the categories, using artifacts provided by the participants through the centralized repository [9]. In the early years, this process was performed by providing installation scripts for the tools as well as execution scripts for the benchmarks, which were then installed on a virtual machine and executed on a laptop, requiring a significant amount of manual effort (around an hour per tool to install and then execute the benchmarks). The process evolved to require Docker files to be submitted so that Docker containers could be built automatically, then with batch execution scripts for all the benchmarks for a given tool in a given category. With the Dockerized setup, the manual intervention typically required around a day of effort to set everything up and start batch execution, with typically around a week of total time required for execution of all benchmarks across all categories. Once standardized with the Dockerized execution, we typically have run our experiments on an Amazon Web Services (AWS) Elastic Compute Cloud (EC2) g4dn.4xlarge machine with an Intel Xeon Scalable (2nd Generation Cascade Lake), 16 vCPUs, 2.5 GHz base, (AWS/EC2 custom chip, roughly Xeon Gold 5200 Series with 24 physical cores) and 64GB RAM. On this platform, the execution runtimes in a given category varied depending on the benchmarks, with some typically completing in seconds with others requiring hours of execution. Further improvements include executing all tools across all the supported benchmarks in each category, along with some collection of performance metrics (runtime, verification result where applicable, etc.).

6 Overall Achievements and Outlook

Since 2017, the series of friendly ARCH-COMP competitions has matured and gained some routine around the yearly schedule. The event enjoys a steady popularity around an active community. New teams with new approaches and tools enter the competition yearly, while categories become more firmly established. For example, for the falsification group, the number of tools participating in the competition (see Table 1) increased from one tool (2017) to eight tools (2021 and 2023). Thanks to theoretical advances and tool improvements, the scale of treatable problems has increased by several orders of magnitude, e.g., in the categories for piecewise continuous, linear, and non-linear categories. In some instances, the insights gained by investigating the results of the competition played an important role.

All groups have worked to increase the number and diversity of the available benchmarks. For example, for the falsification group, the number of benchmark models (see Table 2) increased from one model (2017) to seven models (2020, 2021, and 2023). They are now collected in a central, shared repository for easier access [9]. These benchmarks are regularly used in publications for evaluating new approaches, improving the relevance of such experiments and the comparability across different publications. ARCH-COMP has also gained some visibility in the industry, tools are increasingly applicable to industrial products, and some contributions to the associated workshop are closely related to the competition. For example, recent work [77] enables the use of falsification-based testing techniques with Test Sequence and Test Assessment Blocks, i.e., Simulink blocks commonly used by industrial practitioners to test their models.

Progress has also been made for the concerns of fair evaluation, as discussed in Sect. 2.6: Many groups have established standard formats for problem specifications and exchange formats for witnesses, which increases trust in the results. Similarly, the efforts towards complete repeatability of the evaluation, presented in Sect. 5, are progressing with each installment of ARCH-COMP. Most groups now have automated at least some aspects of this process.

While the competition has seen a lot of progress over the years, some challenges remain. In some categories, fully automating the repeatability evaluation requires other issues to be resolved first, such as standardizing input and witness formats. Due to the diversity and complexity of methods and problems, this, in turn, will require further theoretical and technical advances complemented by implementation work. This is in particular more evident in the Stochastic Category due to the large differences between the class of models and problem formulations on these models. This category is pushing activities on theoretical analysis of stochastic models, developing new tools based on these advances, moving towards an interchange format of stochastic models, and importing minimal case studies where methods developed for different model classes can be compared. With its collaborative spirit, the friendly ARCH-COMP competition will do its best to nurture discussions and collaborations that help to tackle these challenges in the rich and rewarding field of continuous and hybrid systems.

Acknowledgements. We are grateful to all the participants in all the iterations of ARCH-COMP. In addition, we gratefully acknowledge financial support by the project TRAITS, funded by the German Federal Ministry of Education and Research under grant number 01IS21087 and by the French Agence Nationale de Recherche under grant number ANR-21-FAI1-0005-01. Some material presented in this paper is based upon work supported by the National Science Foundation (NSF) through grant numbers 2028001, 2220401, and 2220426, the Defense Advanced Research Projects Agency (DARPA) under contract number FA8750-23-C-0518, and the Air Force Office of Scientific Research (AFOSR) under contract number FA9550-22-1-0019 and FA9550-23-1-0135.

References

1. Abate, A., et al.: Arch-comp21 category report: stochastic models. In: International Workshop on Applied Verification of Continuous and Hybrid Systems (ARCH 2021). EPiC Series in Computing, vol. 80, pp. 55–89. EasyChair (2021). https://doi.org/10.29007/dprv

2. Abate, A., et al.: ARCH-COMP19 category report: stochastic modelling. In: International Workshop on Applied Verification of Continuous and Hybrid Systems (ARCH 2019). EPiC Series in Computing, vol. 61, pp. 62–102. EasyChair (2019). https://doi.org/10.29007/f2vb

3. Abate, A., et al.: Arch-comp20 category report: stochastic models. In: International Workshop on Applied Verification of Continuous and Hybrid Systems (ARCH 2020). EPiC Series in Computing, vol. 74, pp. 76–106. EasyChair (2020). https://doi.org/10.29007/mqzc

4. Abate, A., et al.: ARCH-COMP18 category report: stochastic modelling. In: International Workshop on Applied Verification of Continuous and Hybrid Systems (ARCH 2018). EPiC Series in Computing, vol. 54, pp. 71–103. EasyChair (2018). https://doi.org/10.29007/7ks7

5. Abate, A., et al.: ARCH-COMP22 Category Report: Stochastic Models, vol. 90, pp. 113–141. EasyChair (2022). https://doi.org/10.29007/LSVC

6. Adzkiya, D., Abate, A.: VeriSiMPL: verification via biSimulations of MPL models. In: Joshi, K., Siegle, M., Stoelinga, M., D'Argenio, P.R. (eds.) QEST 2013. LNCS, vol. 8054, pp. 274–277. Springer, Heidelberg (2013). https://doi.org/10.1007/978-3-642-40196-1_22

7. Adzkiya, D., Zhang, Y., Abate, A.: VeriSiMPL 2: an open-source software for the verification of max-plus-linear systems. Discrete Event Dyn. Syst. **26**(1), 109–145 (2016). https://doi.org/10.1007/s10626-015-0218-x

8. Akintunde, M.E., Botoeva, E., Kouvaros, P., Lomuscio, A.: Formal verification of neural agents in non-deterministic environments. In: International Conference on Autonomous Agents and Multiagent Systems, AAMAS, pp. 25–33 (2020)

9. ARCH-COMP repository of benchmark models, documentation, and repeatability packages. https://gitlab.com/goranf/ARCH-COMP

10. Althoff, M.: An introduction to CORA 2015. In: Workshop on Applied Verification for Continuous and Hybrid Systems, pp. 120–151 (2015)
11. Althoff, M.: Reachability analysis of large linear systems with uncertain inputs in the Krylov subspace. IEEE Trans. Autom. Control **65**(2), 477–492 (2020)
12. Althoff, M., et al.: ARCH-COMP20 category report: continuous and hybrid systems with linear continuous dynamics. In: International Workshop on Applied Verification of Continuous and Hybrid Systems. EPiC Series in Computing, vol. 74, pp. 16–48 (2020)
13. Althoff, M., et al.: ARCH-COMP17 category report: continuous and hybrid systems with linear continuous dynamics. In: International Workshop on Applied Verification for Continuous and Hybrid Systems, pp. 143–159 (2017)
14. Althoff, M., et al.: ARCH-COMP18 category report: continuous and hybrid systems with linear continuous dynamics. In: International Workshop on Applied Verification for Continuous and Hybrid Systems, pp. 23–52 (2018)
15. Althoff, M., et al.: ARCH-COMP19 category report: continuous and hybrid systems with linear continuous dynamics. In: International Workshop on Applied Verification of Continuous and Hybrid Systems. EPiC Series in Computing, vol. 61, pp. 14–40 (2019)
16. Althoff, M., Krogh, B.H.: Avoiding geometric intersection operations in reachability analysis of hybrid systems. In: Hybrid Systems: Computation and Control, pp. 45–54 (2012)
17. Althoff, M., et al.: ARCH-COMP21 category report: continuous and hybrid systems with linear continuous dynamics. In: International Workshop on Applied Verification of Continuous and Hybrid Systems, vol. 80, pp. 1–31 (2021). https://doi.org/10.29007/lhbw. https://easychair.org/publications/paper/81BS
18. Althoff, M., Forets, M., Schilling, C., Wetzlinger, M.: ARCH-COMP22 category report: continuous and hybrid systems with linear continuous dynamics. In: International Workshop on Applied Verification of Continuous and Hybrid Systems. EPiC Series in Computing, vol. 90, pp. 58–85. EasyChair (2022). https://doi.org/10.29007/mmzc. https://easychair.org/publications/paper/b6cN
19. Annpureddy, Y., Liu, C., Fainekos, G., Sankaranarayanan, S.: S-TaLiRo: a tool for temporal logic falsification for hybrid systems. In: Abdulla, P.A., Leino, K.R.M. (eds.) TACAS 2011. LNCS, vol. 6605, pp. 254–257. Springer, Heidelberg (2011). https://doi.org/10.1007/978-3-642-19835-9_21
20. Bak, S., Bogomolov, S., Althoff, M.: Time-triggered conversion of guards for reachability analysis of hybrid automata. In: International Conference on Formal Modelling and Analysis of Timed Systems, pp. 133–150 (2017)
21. Bak, S., Bogomolov, S., Johnson, T.T.: HYST: a source transformation and translation tool for hybrid automaton models. In: Proceedings of the 18th International Conference on Hybrid Systems: Computation and Control (2015)
22. Bak, S., Duggirala, P.S.: HyLAA: a tool for computing simulation-equivalent reachability for linear systems. In: Proceedings of the 20th International Conference on Hybrid Systems: Computation and Control, pp. 173–178 (2017)
23. Bak, S., Tran, H.D., Johnson, T.T.: Numerical verification of affine systems with up to a billion dimensions. In: Proceedings of the 22nd ACM International Conference on Hybrid Systems: Computation and Control, pp. 23–32 (2019)
24. Bartocci, E., et al.: Specification-based monitoring of cyber-physical systems: a survey on theory, tools and applications. In: Bartocci, E., Falcone, Y. (eds.) Lectures on Runtime Verification. LNCS, vol. 10457, pp. 135–175. Springer, Cham (2018). https://doi.org/10.1007/978-3-319-75632-5_5

25. Becchi, A., Zaffanella, E.: A direct encoding for NNC Polyhedra. In: Chockler, H., Weissenbacher, G. (eds.) CAV 2018. LNCS, vol. 10981, pp. 230–248. Springer, Cham (2018). https://doi.org/10.1007/978-3-319-96145-3_13
26. van Beek, D.A., Reniers, M.A., Rooda, J.E., Schiffelers, R.R.: Concrete syntax and semantics of the compositional interchange format for hybrid systems. IFAC Proc. Vol. **41**(2), 7979–7986 (2008)
27. Bemporad, A.: Efficient conversion of mixed logical dynamical systems into an equivalent piecewise affine form. IEEE Trans. Autom. Control **49**(5), 832–838 (2004)
28. Berz, M., Makino, K.: Performance of Taylor model methods for validated integration of ODEs. In: Dongarra, J., Madsen, K., Waśniewski, J. (eds.) PARA 2004. LNCS, vol. 3732, pp. 65–73. Springer, Heidelberg (2006). https://doi.org/10.1007/11558958_8
29. Beyer, D.: Competition on software verification and witness validation: SV-COMP 2023. In: Sankaranarayanan, S., Sharygina, N. (eds.) 13994. LNCS, vol. 13994, pp. 495–522. Springer, Cham (2023). https://doi.org/10.1007/978-3-031-30820-8_29
30. Blom, H.A., Ma, H., Bakker, G.B.: Interacting particle system-based estimation of reach probability for a generalized stochastic hybrid system. IFAC-PapersOnLine **51**(16), 79–84 (2018)
31. Bogomolov, S., Forets, M., Frehse, G., Potomkin, K., Schilling, C.: JuliaReach: a toolbox for set-based reachability. In: Proceedings of the 22nd ACM International Conference on Hybrid Systems: Computation and Control, pp. 39–44 (2019). https://doi.org/10.1145/3302504.3311804
32. Bogomolov, S., Forets, M., Frehse, G., Viry, F., Podelski, A., Schilling, C.: Reach set approximation through decomposition with low-dimensional sets and high-dimensional matrices. In: Proceedings of the 21st International Conference on Hybrid Systems: Computation and Control, pp. 41–50 (2018)
33. Bogomolov, S., Frehse, G., Giacobbe, M., Henzinger, T.A.: Counterexample-guided refinement of template polyhedra. In: Legay, A., Margaria, T. (eds.) TACAS 2017. LNCS, vol. 10205, pp. 589–606. Springer, Heidelberg (2017). https://doi.org/10.1007/978-3-662-54577-5_34
34. Bohrer, R., Rahli, V., Vukotic, I., Völp, M., Platzer, A.: Formally verified differential dynamic logic. In: CPP, pp. 208–221. ACM (2017)
35. Bohrer, R., Tan, Y.K., Mitsch, S., Myreen, M.O., Platzer, A.: Veriphy: verified controller executables from verified cyber-physical system models. In: ACM SIGPLAN Conference on Programming Language Design and Implementation, PLDI 2018, pp. 617–630 (2018). https://doi.org/10.1145/3192366.3192406
36. Bouissou, M., Houdebine, J.: Inconsistency detection in KB3 models. In: ESREL 2002 (2002)
37. Bouissou, M., Houdebine, S., Houdebine, J.C.: Reference manual of the Figaro probabilistic modelling language (2019)
38. Bouissou, M., Khan, S.: Bridging the dependability and model checking worlds. In: Congrès Lambda Mu 23 «Innovations et maîtrise des risques pour un avenir durable»-23e Congrès de Maîtrise des Risques et de Sûreté de Fonctionnement, Institut pour la Maîtrise des Risques (2022)
39. Bourke, T., Pouzet, M.: Zélus: a synchronous language with ODEs. In: International Conference on Hybrid Systems: Computation and Control (HSCC), pp. 113–118 (2013)
40. Bresolin, D., Collins, P., Geretti, L., Segala, R., Villa, T., Gonzalez, S.V.: A computable and compositional semantics for hybrid automata. In: International Conference on Hybrid Systems: Computation and Control HSCC. ACM (2020)

41. Bu, L., et al.: ARCH-COMP20 category report: hybrid systems with piecewise constant dynamics and bounded model checking. In: International Workshop on Applied Verification of Continuous and Hybrid Systems (ARCH20). EPiC Series in Computing, vol. 74, pp. 1–15. EasyChair (2020)

42. Bu, L., Frehse, G., Kundu, A., Ray, R., Shi, Y., Zaffanella, E.: ARCH-COMP22 category report: hybrid systems with piecewise constant dynamics and bounded model checking. In: International Workshop on Applied Verification of Continuous and Hybrid Systems. EPiC Series in Computing, vol. 90, pp. 44–57. EasyChair (2022)

43. Bu, L., Li, Y., Wang, L., Chen, X., Li, X.: BACH 2: bounded reachability checker for compositional linear hybrid systems. In: Design, Automation and Test in Europe (DATE), pp. 1512–1517 (2010)

44. Bu, L., Li, Y., Wang, L., Li, X.: BACH: bounded reachability checker for linear hybrid automata. In: Formal Methods in Computer-Aided Design (FMCAD), pp. 1–4 (2008)

45. Bu, L., Ray, R., Schupp, S.: ARCH-COMP17 category report: bounded model checking of hybrid systems with piecewise constant dynamics. In: ARCH 2017. International Workshop on Applied Verification of Continuous and Hybrid Systems, collocated with Cyber-Physical Systems Week (CPSWeek). EPiC Series in Computing, vol. 48, pp. 134–142. EasyChair (2017)

46. Bu, L., Ray, R., Schupp, S.: ARCH-COMP18 category report: bounded model checking of hybrid systems with piecewise constant dynamics. In: ARCH 2018. International Workshop on Applied Verification of Continuous and Hybrid Systems. EPiC Series in Computing, vol. 54, pp. 14–22. EasyChair (2018)

47. Bu, L., Ray, R., Schupp, S.: ARCH-COMP19 category report: bounded model checking of hybrid systems with piecewise constant dynamics. In: ARCH 2019. International Workshop on Applied Verification of Continuous and Hybrid Systems, part of CPS-IoT Week. EPiC Series in Computing, vol. 61, pp. 120–128. EasyChair (2019)

48. Budde, C.E., D'Argenio, P.R., Hartmanns, A., Sedwards, S.: An efficient statistical model checker for nondeterminism and rare events. Int. J. Softw. Tools Technol. Transfer **22**(6), 759–780 (2020)

49. Cattaruzza, D., Abate, A., Schrammel, P., Kroening, D.: Unbounded-time analysis of guarded LTI systems with inputs by abstract acceleration. In: Static Analysis, pp. 312–331 (2015)

50. Cauchi, N., Abate, A.: StocHy: automated verification and synthesis of stochastic processes. In: International Conference on Tools and Algorithms for the Construction and Analysis of Systems (TACAS) (2019)

51. Chen, X., Ábrahám, E., Sankaranarayanan, S.: Flow*: an analyzer for non-linear hybrid systems. In: Sharygina, N., Veith, H. (eds.) CAV 2013. LNCS, vol. 8044, pp. 258–263. Springer, Heidelberg (2013). https://doi.org/10.1007/978-3-642-39799-8_18

52. Chen, X., Althoff, M., Immler, F.: Arch-comp17 category report: continuous systems with nonlinear dynamics. In: International Workshop on Applied Verification of Continuous and Hybrid Systems. EPiC Series in Computing, vol. 48, pp. 160–169. EasyChair (2017). https://doi.org/10.29007/v6g4

53. Chraibi, H., Houbedine, J., Sibler, A.: PyCATSHOO: toward a new platform dedicated to dynamic reliability assessments of hybrid systems. In: 13th International Conference on Probabilistic Safety Assessment and Management (PSAM 13), Seoul, Korea (2016)

54. Chraibi, H., Houbedine, J., Sibler, A.: Pycatshoo: toward a new platform dedicated to dynamic reliability assessments of hybrid systems. In: PSAM Congress (2016)
55. Cimatti, A., Griggio, A., Mover, S., Tonetta, S.: HyComp: an SMT-based model checker for hybrid systems. In: TACAS, pp. 52–67 (2015)
56. Corso, A., Moss, R.J., Koren, M., Lee, R., Kochenderfer, M.J.: A survey of algorithms for black-box safety validation of cyber-physical systems. J. Artif. Intell. Res. **72**, 377–428 (2021). https://doi.org/10.1613/jair.1.12716
57. Dang, T., Testylier, R.: Reachability analysis for polynomial dynamical systems using the Bernstein expansion. Reliable Comput. **17** (2012)
58. Delicaris, J., Schupp, S., Ábrahám, E., Remke, A.: Maximizing reachability probabilities in rectangular automata with random clocks. In: David, C., Sun, M. (eds.) TASE 2023. LNCS, vol. 13931. Springer, Cham (2023). https://doi.org/10.1007/978-3-031-35257-7_10
59. Dokhanchi, A., Yaghoubi, S., Hoxha, B., Fainekos, G.: ARCH-COMP17 category report: preliminary results on the falsification benchmarks. In: ARCH 2017. International Workshop on Applied Verification of Continuous and Hybrid Systems. EPiC Series in Computing, pp. 170–174. EasyChair (2017). https://doi.org/10.29007/wmf5
60. Dokhanchi, A., et al.: ARCH-COMP18 category report: results on the falsification benchmarks. In: ARCH 2018. International Workshop on Applied Verification of Continuous and Hybrid Systems. EPiC Series in Computing, pp. 104–109. EasyChair (2018). https://doi.org/10.29007/t85q
61. Donzé, A.: Breach, a toolbox for verification and parameter synthesis of hybrid systems. In: Proceedings of Computer-Aided Verification, pp. 167–170 (2010)
62. Duggirala, P.S., Mitra, S., Viswanathan, M., Potok, M.: C2E2: a verification tool for stateflow models. In: Tools and Algorithms for the Construction and Analysis of Systems, pp. 68–82 (2015)
63. Dutta, S., Chen, X., Sankaranarayanan, S.: Reachability analysis for neural feedback systems using regressive polynomial rule inference. In: ACM International Conference on Hybrid Systems: Computation and Control, HSCC, pp. 157–168 (2019). https://doi.org/10.1145/3302504.3311807
64. Dutta, S., Jha, S., Sankaranarayanan, S., Tiwari, A.: Learning and verification of feedback control systems using feedforward neural networks. IFAC-PapersOnLine **51**(16), 151–156 (2018). https://doi.org/10.1016/j.ifacol.2018.08.026. iFAC Conference on Analysis and Design of Hybrid Systems ADHS 2018
65. Dutta, S., Jha, S., Sankaranarayanan, S., Tiwari, A.: Output range analysis for deep feedforward neural networks. In: Dutle, A., Muñoz, C., Narkawicz, A. (eds.) NFM 2018. LNCS, vol. 10811, pp. 121–138. Springer, Cham (2018). https://doi.org/10.1007/978-3-319-77935-5_9
66. Ernst, G., et al.: ARCH-COMP 2021 category report: falsification with validation of results. In: International Workshop on Applied Verification of Continuous and Hybrid Systems (ARCH 2021). EPiC Series in Computing, pp. 133–152. EasyChair (2021). https://doi.org/10.29007/xwll
67. Ernst, G., et al.: ARCH-COMP 2020 category report: falsification. In: ARCH 2020. International Workshop on Applied Verification of Continuous and Hybrid Systems (ARCH 2020). EPiC Series in Computing, pp. 140–152. EasyChair (2020). https://doi.org/10.29007/trr1

68. Ernst, G., et al.: ARCH-COMP 2019 category report: falsification. In: ARCH 2019. International Workshop on Applied Verification of Continuous and Hybrid Systems. EPiC Series in Computing, pp. 129–140. EasyChair (2019). https://doi.org/10.29007/68dk

69. Ernst, G., et al.: Arch-comp 2022 category report: falsification with ubounded resources. In: International Workshop on Applied Verification of Continuous and Hybrid Systems (ARCH 2022). EPiC Series in Computing, pp. 204–221. Easy-Chair (2022). https://doi.org/10.29007/fhnk

70. Ernst, G., Sedwards, S., Zhang, Z., Hasuo, I.: Falsification of hybrid systems using adaptive probabilistic search. ACM Trans. Model. Comput. Simul. (TOMACS) **31**(3), 1–22 (2021)

71. Everdij, M., Blom, H.: Hybrid state Petri nets which have the analysis power of stochastic hybrid systems and the formal verification power of automata. In: Pawlewski, P. (ed.) Petri Nets, chap. 12, pp. 227–252. I-Tech Education and Publishing, Vienna (2010)

72. Immler, F., Althoff, M., et al.: Symreach. https://github.com/mahendrasinghtomar/SymReach

73. Fainekos, G.E., Pappas, G.J.: Robustness of temporal logic specifications. In: Havelund, K., Núñez, M., Roşu, G., Wolff, B. (eds.) FATES/RV -2006. LNCS, vol. 4262, pp. 178–192. Springer, Heidelberg (2006). https://doi.org/10.1007/11940197_12

74. Fan, C., Qi, B., Mitra, S., Viswanathan, M., Duggirala, P.S.: Automatic reachability analysis for nonlinear hybrid models with C2E2. In: Computer Aided Verification, pp. 531–538 (2016)

75. Fan, J., Huang, C., Li, W., Chen, X., Zhu, Q.: Reachnn*: a tool for reachability analysis of neural-network controlled systems. In: International Symposium on Automated Technology for Verification and Analysis (ATVA) (2020)

76. Fijalkow, N., Ouaknine, J., Pouly, A., Sousa-Pinto, J., Worrell, J.: On the decidability of reachability in linear time-invariant systems. In: Proceedings of the 22nd ACM International Conference on Hybrid Systems: Computation and Control, pp. 77–86 (2019)

77. Formica, F., Fan, T., Rajhans, A., Pantelic, V., Lawford, M., Menghi, C.: Simulation-based testing of simulink models with test sequence and test assessment blocks. IEEE Trans. Softw. Eng. 1–19 (2023). https://doi.org/10.1109/TSE.2023.3343753

78. Formica, F., Tony, F., Menghi, C.: Search-based software testing driven by automatically generated and manually defined fitness functions. ACM Trans. Softw. Eng. Methodol. (2023). https://doi.org/10.1145/3624745

79. Foster, S., Huerta y Munive, J.J., Gleirscher, M., Struth, G.: Hybrid systems verification with Isabelle/HOL: simpler syntax, better models, faster proofs. In: Huisman, M., Păsăreanu, C., Zhan, N. (eds.) FM 2021. LNCS, vol. 13047, pp. 367–386. Springer, Cham (2021). https://doi.org/10.1007/978-3-030-90870-6_20

80. Fränzle, M., Hahn, E.M., Hermanns, H., Wolovick, N., Zhang, L.: Measurability and safety verification for stochastic hybrid systems. In: International Conference on Hybrid Systems: Computation and Control, HSCC 2011, pp. 43–52. ACM (2011). https://doi.org/10.1145/1967701.1967710

81. Fränzle, M., Hansen, M.R.: A robust interpretation of duration calculus. In: Van Hung, D., Wirsing, M. (eds.) ICTAC 2005. LNCS, vol. 3722, pp. 257–271. Springer, Heidelberg (2005). https://doi.org/10.1007/11560647_17

82. Frehse, G.: PHAVer: algorithmic verification of hybrid systems past HyTech. Int. J. Softw. Tools Technol. Transfer **10**, 263–279 (2008)

83. Frehse, G., et al.: SpaceEx: scalable verification of hybrid systems. In: Gopalakrishnan, G., Qadeer, S. (eds.) CAV 2011. LNCS, vol. 6806, pp. 379–395. Springer, Heidelberg (2011). https://doi.org/10.1007/978-3-642-22110-1_30

84. Frehse, G., et al.: ARCH-COMP19 category report: hybrid systems with piecewise constant dynamics. In: ARCH 2019. International Workshop on Applied Verification of Continuous and Hybrid Systems, part of CPS-IoT Week. EPiC Series in Computing, vol. 61, pp. 1–13. EasyChair (2019)

85. Frehse, G., Abate, A., Adzkiya, D., Bu, L., Giacobbe, M.: ARCH-COMP17 category report: hybrid systems with piecewise constant dynamics. In: ARCH 2017. International Workshop on Applied Verification of Continuous and Hybrid Systems, collocated with Cyber-Physical Systems Week (CPSWeek). EPiC Series in Computing, vol. 48, pp. 124–133. EasyChair (2017)

86. Frehse, G., et al.: ARCH-COMP18 category report: hybrid systems with piecewise constant dynamics. In: ARCH 2018. International Workshop on Applied Verification of Continuous and Hybrid Systems, ARCH@ADHS. EPiC Series in Computing, vol. 54, pp. 1–13. EasyChair (2018)

87. Frehse, G.F.: Compositional verification of hybrid systems using simulation relations. Ph.D. thesis, Radboud University (2005)

88. Fulton, N., Mitsch, S., Bohrer, B., Platzer, A.: Bellerophon: tactical theorem proving for hybrid systems. In: Ayala-Rincón, M., Muñoz, C.A. (eds.) ITP 2017. LNCS, vol. 10499, pp. 207–224. Springer, Cham (2017). https://doi.org/10.1007/978-3-319-66107-0_14

89. Fulton, N., Mitsch, S., Quesel, J.-D., Völp, M., Platzer, A.: KeYmaera X: an axiomatic tactical theorem prover for hybrid systems. In: Felty, A.P., Middeldorp, A. (eds.) CADE 2015. LNCS (LNAI), vol. 9195, pp. 527–538. Springer, Cham (2015). https://doi.org/10.1007/978-3-319-21401-6_36

90. Fulton, N., Platzer, A.: Safe reinforcement learning via formal methods: toward safe control through proof and learning. In: Conference on Artificial Intelligence, (AAAI), pp. 6485–6492 (2018)

91. Gallicchio, J., Tan, Y.K., Mitsch, S., Platzer, A.: Implicit definitions with differential equations for KeYmaera X - (system description). In: Blanchette, J., Kovács, L., Pattinson, D. (eds.) IJCAR 2022. LNCS, vol. 13385, pp. 723–733. Springer, Cham (2022). https://doi.org/10.1007/978-3-031-10769-6_42

92. Garcia, L., Mitsch, S., Platzer, A.: Hyplc: hybrid programmable logic controller program translation for verification. In: ACM/IEEE International Conference on Cyber-Physical Systems, ICCPS, pp. 47–56 (2019). https://doi.org/10.1145/3302509.3311036

93. Geretti, L., Collins, P., Bresolin, D., Villa, T.: Automating numerical parameters along the evolution of a nonlinear system. In: Dang, T., Stolz, V. (eds.) RV 2022. LNCS, vol. 13498, pp. 336–345. Springer, Cham (2022). https://doi.org/10.1007/978-3-031-17196-3_22

94. Geretti, L., et al.: Arch-comp20 category report: continuous and hybrid systems with nonlinear dynamics. In: International Workshop on Applied Verification of Continuous and Hybrid Systems (ARCH20). EPiC Series in Computing, vol. 74, pp. 49–75. EasyChair (2020). https://doi.org/10.29007/zkf6. https://easychair.org/publications/paper/nrdD

95. Geretti, L., et al.: ARCH-COMP21 category report: continuous and hybrid systems with nonlinear dynamics. In: International Workshop on Applied Verification of Continuous and Hybrid Systems (ARCH21). EPiC Series in Computing, vol. 80, pp. 32–54. EasyChair (2021). https://doi.org/10.29007/2jw8. https://easychair.org/publications/paper/GWwz

96. Geretti, L., et al.: ARCH-COMP22 category report: continuous and hybrid systems with nonlinear dynamics. In: International Workshop on Applied Verification of Continuous and Hybrid Systems (ARCH22). EPiC Series in Computing, vol. 90, pp. 86–112. EasyChair (2022). https://doi.org/10.29007/fnzc. https://easychair.org/publications/paper/JrQ4

97. Geretti, L., et al.: Arch-comp23 category report: continuous and hybrid systems with nonlinear dynamics. In: Frehse, G., Althoff, M. (eds.) Proceedings of 10th International Workshop on Applied Verification of Continuous and Hybrid Systems (ARCH23). EPiC Series in Computing, vol. 96, pp. 61–88. EasyChair (2023). https://doi.org/10.29007/93f2. https://easychair.org/publications/paper/T7LG

98. Girard, A., Le Guernic, C.: Zonotope/hyperplane intersection for hybrid systems reachability analysis. In: Egerstedt, M., Mishra, B. (eds.) HSCC 2008. LNCS, vol. 4981, pp. 215–228. Springer, Heidelberg (2008). https://doi.org/10.1007/978-3-540-78929-1_16

99. Girard, A., Le Guernic, C., Maler, O.: Efficient computation of reachable sets of linear time-invariant systems with inputs. In: Hespanha, J.P., Tiwari, A. (eds.) HSCC 2006. LNCS, vol. 3927, pp. 257–271. Springer, Heidelberg (2006). https://doi.org/10.1007/11730637_21

100. Guelev, D.P., Wang, S., Zhan, N.: Compositional hoare-style reasoning about hybrid CSP in the duration calculus. In: Larsen, K.G., Sokolsky, O., Wang, J. (eds.) SETTA 2017. LNCS, vol. 10606, pp. 110–127. Springer, Cham (2017). https://doi.org/10.1007/978-3-319-69483-2_7

101. Haesaert, S., Soudjani, S.: Robust dynamic programming for temporal logic control of stochastic systems. IEEE Trans. Autom. Control **66**(6), 2496–2511 (2020)

102. Haesaert, S., Zadeh Soudjani, S.E., Abate, A.: Verification of general Markov decision processes by approximate similarity relations and policy refinement. SIAM J. Control. Optim. **55**(4), 2333–2367 (2017)

103. Hartmanns, A., Hermanns, H.: The modest toolset: an integrated environment for quantitative modelling and verification. In: Ábrahám, E., Havelund, K. (eds.) TACAS 2014. LNCS, vol. 8413, pp. 593–598. Springer, Heidelberg (2014). https://doi.org/10.1007/978-3-642-54862-8_51

104. Henzinger, T.: The theory of hybrid automata. In: Inan, K., Kurshan, R.P. (eds.) Verification of Digital and Hybrid Systems. NATO ASI Series, vol. 170, pp. 265–292. Springer, Heidelberg (2000). https://doi.org/10.1007/978-3-642-59615-5_13

105. Huang, C., Fan, J., Chen, X., Li, W., Zhu, Q.: POLAR: a polynomial arithmetic framework for verifying neural-network controlled systems. In: International Symposium on Automated Technology for Verification and Analysis (ATVA) (2022)

106. Huang, C., Fan, J., Li, W., Chen, X., Zhu, Q.: Reachnn: reachability analysis of neural-network controlled systems. ACM Trans. Embed. Comput. Syst. (TECS) **18**(5s), 1–22 (2019)

107. Huerta y Munive, J.J., Struth, G.: Predicate transformer semantics for hybrid systems. JAR **66**(1), 93–139 (2022)

108. Hüls, J., Niehaus, H., Remke, A.: hpnmg: a C++ tool for model checking hybrid petri nets with general transitions. In: Lee, R., Jha, S., Mavridou, A., Giannakopoulou, D. (eds.) NFM 2020. LNCS, vol. 12229, pp. 369–378. Springer, Cham (2020). https://doi.org/10.1007/978-3-030-55754-6_22

109. Immler, F.: Verified reachability analysis of continuous systems. In: Baier, C., Tinelli, C. (eds.) TACAS 2015. LNCS, vol. 9035, pp. 37–51. Springer, Heidelberg (2015). https://doi.org/10.1007/978-3-662-46681-0_3

110. Immler, F., et al.: ARCH-COMP19 category report: continuous and hybrid systems with nonlinear dynamics. In: ARCH 2019. International Workshop on Applied Verification of Continuous and Hybrid Systems. EPiC Series in Computing, vol. 61, pp. 41–61. EasyChair (2019). https://doi.org/10.29007/m75b. https://easychair.org/publications/paper/4FSh

111. Immler, F., et al.: ARCH-COMP18 category report: continuous and hybrid systems with nonlinear dynamics. In: Frehse, G. (ed.) ARCH 2018. International Workshop on Applied Verification of Continuous and Hybrid Systems. EPiC Series in Computing, vol. 54, pp. 53–70. EasyChair (2018).https://doi.org/10.29007/mskf. https://easychair.org/publications/paper/gjfh

112. Ivanov, R., Carpenter, T., Weimer, J., Alur, R., Pappas, G.J., Lee, I.: Verisig 2.0: verification of neural network controllers using taylor model preconditioning. In: International Conference on Computer-Aided Verification (2021)

113. Ivanov, R., Weimer, J., Alur, R., Pappas, G.J., Lee, I.: Verisig: verifying safety properties of hybrid systems with neural network controllers. In: International Conference on Hybrid Systems: Computation and Control, HSCC, pp. 169–178. ACM (2019). https://doi.org/10.1145/3302504.3311806

114. Jagtap, P., Soudjani, S., Zamani, M.: Temporal logic verification of stochastic systems using barrier certificates. In: Lahiri, S.K., Wang, C. (eds.) ATVA 2018. LNCS, vol. 11138, pp. 177–193. Springer, Cham (2018). https://doi.org/10.1007/978-3-030-01090-4_11

115. Jin, X., Deshmukh, J.V., Kapinski, J., Ueda, K., Butts, K.: Powertrain control verification benchmark. In: International Conference on Hybrid Systems: Computation and Control, pp. 253–262. ACM (2014)

116. Johnson, T.T.: Arch-comp17 repeatability evaluation report. In: ARCH 2017. International Workshop on Applied Verification of Continuous and Hybrid Systems. EPiC Series in Computing, vol. 48, pp. 175–180. EasyChair (2017). https://doi.org/10.29007/7hvk. https://easychair.org/publications/paper/nMvb

117. Johnson, T.T.: Arch-comp18 repeatability evaluation report. In: ARCH 2018. International Workshop on Applied Verification of Continuous and Hybrid Systems. EPiC Series in Computing, vol. 54, pp. 128–134. EasyChair (2018). https://doi.org/10.29007/n9t3. https://easychair.org/publications/paper/9J6v

118. Johnson, T.T.: Arch-comp19 repeatability evaluation report. In: ARCH 2019. International Workshop on Applied Verification of Continuous and Hybrid Systems. EPiC Series in Computing, vol. 61, pp. 162–169. EasyChair (2019).https://doi.org/10.29007/wbl3. https://easychair.org/publications/paper/xvBM

119. Johnson, T.T.: Arch-comp20 repeatability evaluation report. In: ARCH 2020. International Workshop on Applied Verification of Continuous and Hybrid Systems (ARCH20). EPiC Series in Computing, vol. 74, pp. 175–183. EasyChair (2020). https://doi.org/10.29007/8dp4. https://easychair.org/publications/paper/3W11

120. Johnson, T.T.: ARCH-COMP 2021 repeatability evaluation report. In: International Workshop on Applied Verification of Continuous and Hybrid Systems (ARCH21). EPiC Series in Computing, vol. 80, pp. 153–160. EasyChair (2021). https://doi.org/10.29007/zqdx. https://easychair.org/publications/paper/cfpN

121. Johnson, T.T.: Arch-comp22 repeatability evaluation report. In: International Workshop on Applied Verification of Continuous and Hybrid Systems (ARCH 2022). EPiC Series in Computing, vol. 90, pp. 222–230. EasyChair (2022). https://doi.org/10.29007/djqx. https://easychair.org/publications/paper/LnDH

122. Johnson, T.T.: Arch-comp23 repeatability evaluation report. In: Frehse, G., Althoff, M. (eds.) Proceedings of 10th International Workshop on Applied Verification of Continuous and Hybrid Systems (ARCH 2023). EPiC Series in Computing, vol. 96, pp. 189–195. EasyChair (2023). https://doi.org/10.29007/q313. https://easychair.org/publications/paper/TdVx

123. Johnson, T.T., et al.: Arch-comp21 category report: artificial intelligence and neural network control systems (AINNCS) for continuous and hybrid systems plants. In: International Workshop on Applied Verification of Continuous and Hybrid Systems (ARCH 2021). EPiC Series in Computing, vol. 80, pp. 90–119. EasyChair (2021). https://doi.org/10.29007/kfk9

124. Johnson, T.T., et al.: Arch-comp20 category report: artificial intelligence and neural network control systems (AINNCS) for continuous and hybrid systems plants. In: ARCH 2020. International Workshop on Applied Verification of Continuous and Hybrid Systems (ARCH 2020). EPiC Series in Computing, vol. 74, pp. 107–139. EasyChair (2020). https://doi.org/10.29007/9xgv

125. Kim, E., Duggirala, P.S.: Kaa: a python implementation of reachable set computation using bernstein polynomials. EPiC Ser. Comput. **74**, 184–196 (2020)

126. Kochdumper, N., Schilling, C., Althoff, M., Bak, S.: Open- and closed-loop neural network verification using polynomial zonotopes. In: Rozier, K.Y., Chaudhuri, S. (eds.) NFM 2023. LNCS, vol. 13903, pp. 16–36. Springer, Cham (2023). https://doi.org/10.1007/978-3-031-33170-1_2

127. Koymans, R.: Specifying real-time properties with metric temporal logic. Real-Time Syst. **2**(4), 255–299 (1990)

128. Lafferriere, G., Pappas, G.J., Yovine, S.: Symbolic reachability computation for families of linear vector fields. Symb. Comput. **32**, 231–253 (2001)

129. Lavaei, A., Khaled, M., Soudjani, S., Zamani, M.: AMYTISS: parallelized automated controller synthesis for large-scale stochastic systems. In: Lahiri, S.K., Wang, C. (eds.) CAV 2020. LNCS, vol. 12225, pp. 461–474. Springer, Cham (2020). https://doi.org/10.1007/978-3-030-53291-8_24

130. Lavaei, A., Soudjani, S., Abate, A., Zamani, M.: Automated verification and synthesis of stochastic hybrid systems: a survey. Automatica **146**, 110617 (2022)

131. Leahy, K., et al.: Control in belief space with temporal logic specifications using vision-based localization. Int. J. Robot. Res. **38**(6), 702–722 (2019)

132. Li, Y., Zhu, H., Braught, K., Shen, K., Mitra, S.: Verse: a python library for reasoning about multi-agent hybrid system scenarios. In: Computer Aided Verification, pp. 351–364 (2023)

133. Loos, S.M., Platzer, A.: Differential refinement logic. In: Annual ACM/IEEE Symposium on Logic in Computer Science, LICS, pp. 505–514 (2016). https://doi.org/10.1145/2933575.2934555

134. Lopez, D.M., et al.: Arch-comp22 category report: artificial intelligence and neural network control systems (AINNCS) for continuous and hybrid systems plants. In: International Workshop on Applied Verification of Continuous and Hybrid Systems (ARCH 2022). EPiC Series in Computing, vol. 90, pp. 142–184. EasyChair (2022). https://doi.org/10.29007/wfgr

135. Lopez, D.M., Althoff, M., Forets, M., Johnson, T.T., Ladner, T., Schilling, C.: Arch-comp23 category report: artificial intelligence and neural network control systems (AINNCS) for continuous and hybrid systems plants. In: Frehse, G., Althoff, M. (eds.) Proceedings of 10th International Workshop on Applied Verification of Continuous and Hybrid Systems (ARCH 2023). EPiC Series in Computing, vol. 96, pp. 89–125. EasyChair (2023). https://doi.org/10.29007/x38n. https://easychair.org/publications/paper/Vfq4b

136. Lopez, D.M., Choi, S.W., Tran, H.D., Johnson, T.T.: NNV 2.0: the neural network verification tool. In: Enea, C., Lal, A. (eds.) CAV 2023. LNCS, vol. 13965, pp. 397–412. Springer, Heidelberg (2023). https://doi.org/10.1007/978-3-031-37703-7_19

137. Lopez, D.M., et al.: Arch-comp19 category report: artificial intelligence and neural network control systems (AINNCS) for continuous and hybrid systems plants. In: ARCH 2019. International Workshop on Applied Verification of Continuous and Hybrid Systems. EPiC Series in Computing, vol. 61, pp. 103–119. EasyChair (2019). https://doi.org/10.29007/rgv8

138. Ma, H., Blom, H.A.: Interacting particle system based estimation of reach probability of general stochastic hybrid systems. Nonlinear Anal. Hybrid Syst **47**, 101303 (2023)

139. Majumdar, R., Mallik, K., Rychlicki, M., Schmuck, A.K., Soudjani, S.: A flexible toolchain for symbolic rabin games under fair and stochastic uncertainties. In: Enea, C., Lal, A. (eds.) CAV 2023. LNCS, vol. 13966. Springer, Cham (2023). https://doi.org/10.1007/978-3-031-37709-9_1

140. Majumdar, R., Mallik, K., Soudjani, S.: Symbolic controller synthesis for büchi specifications on stochastic systems. In: International Conference on Hybrid Systems: Computation and Control, pp. 1–11 (2020)

141. Maler, O., Nickovic, D.: Monitoring temporal properties of continuous signals. In: Lakhnech, Y., Yovine, S. (eds.) FORMATS/FTRTFT -2004. LNCS, vol. 3253, pp. 152–166. Springer, Heidelberg (2004). https://doi.org/10.1007/978-3-540-30206-3_12

142. Menghi, C., Nejati, S., Briand, L., Isasi Parache, Y.: Approximation-refinement testing of compute-intensive cyber-physical models: an approach based on system identification. In: International Conference on Software Engineering (ICSE). IEEE/ACM (2020)

143. Mitsch, S.: Implicit and explicit proof management in keymaera X. In: Proceedings of the 6th Workshop on Formal Integrated Development Environment, F-IDE@NFM 2021, Held online, 24–25 May 2021, pp. 53–67 (2021). https://doi.org/10.4204/EPTCS.338.8

144. Mitsch, S., Jin, X., Zhan, B., Wang, S., Zhan, N.: Arch-comp21 category report: hybrid systems theorem proving. In: Frehse, G., Althoff, M. (eds.) 8th International Workshop on Applied Verification of Continuous and Hybrid Systems (ARCH21). EPiC Series in Computing, vol. 80, pp. 120–132. EasyChair (2021). https://doi.org/10.29007/35cf

145. Mitsch, S., y Munive, J.J.H., Jin, X., Zhan, B., Wang, S., Zhan, N.: ARCH-COMP20 category report: hybrid systems theorem proving. In: ARCH. EPiC Series in Computing, vol. 74, pp. 153–174. EasyChair (2020)

146. Mitsch, S., Platzer, A.: Modelplex: verified runtime validation of verified cyber-physical system models. Formal Methods Syst. Des. **49**(1–2), 33–74 (2016). https://doi.org/10.1007/s10703-016-0241-z

147. Mitsch, S., Sheng, H., Zhan, B., Wang, S., Foster, S., Munive, J.J.H.Y.: Arch-comp23 category report: hybrid systems theorem proving. In: Frehse, G., Althoff, M. (eds.) Proceedings of 10th International Workshop on Applied Verification of Continuous and Hybrid Systems (ARCH23). EPiC Series in Computing, vol. 96, pp. 170–188. EasyChair (2023). https://doi.org/10.29007/57g4

148. Mitsch, S., et al.: Arch-comp22 category report: hybrid systems theorem proving. In: Frehse, G., Althoff, M., Schoitsch, E., Guiochet, J. (eds.) Proceedings of 9th International Workshop on Applied Verification of Continuous and Hybrid Systems (ARCH 2022). EPiC Series in Computing, vol. 90, pp. 185–203. EasyChair (2022). https://doi.org/10.29007/4lxf

149. Mufid, M.S., Adzkiya, D., Abate, A.: Symbolic reachability analysis of high dimensional max-plus linear systems. IFAC-PapersOnLine 53(4), 459–465 (2020). https://doi.org/10.1016/j.ifacol.2021.04.060

150. y Munive, J.J.H.: Verification components for hybrid systems. Arch. Formal Proofs 2019 (2019). https://www.isa-afp.org/entries/Hybrid_Systems_VCs.html

151. NNFal (2023). https://gitlab.com/Atanukundu/NNFal

152. Peltomäki, J., Porres, I.: Requirement falsification for cyber-physical systems using generative models. arXiv preprint arXiv:2310.20493 (2023)

153. Pilch, C., Remke, A.: HYPEG: statistical model checking for hybrid petri nets: tool paper. In: International Conference on Performance Evaluation Methodologies and Tools, VALUETOOLS 2017, pp. 186–191. ACM (2017)

154. Platzer, A.: A complete uniform substitution calculus for differential dynamic logic. J. Autom. Reason. 59(2), 219–265 (2017)

155. Platzer, A., Quesel, J.-D.: KeYmaera: a hybrid theorem prover for hybrid systems (system description). In: Armando, A., Baumgartner, P., Dowek, G. (eds.) IJCAR 2008. LNCS (LNAI), vol. 5195, pp. 171–178. Springer, Heidelberg (2008). https://doi.org/10.1007/978-3-540-71070-7_15

156. Qian, M., Mitsch, S.: Reward shaping from hybrid systems models in reinforcement learning. In: NASA Formal Methods - International Symposium (NFM), pp. 122–139 (2023). https://doi.org/10.1007/978-3-031-33170-1_8

157. Quesel, J., Mitsch, S., Loos, S.M., Aréchiga, N., Platzer, A.: How to model and prove hybrid systems with keymaera: a tutorial on safety. Int. J. Softw. Tools Technol. Transf. 18(1), 67–91 (2016)

158. Ray, R., Gurung, A., Das, B., Bartocci, E., Bogomolov, S., Grosu, R.: XSpeed: accelerating reachability analysis on multi-core processors. In: Piterman, N. (ed.) HVC 2015. LNCS, vol. 9434, pp. 3–18. Springer, Cham (2015). https://doi.org/10.1007/978-3-319-26287-1_1

159. Salamati, M., Soudjani, S., Majumdar, R.: Approximate time bounded reachability for CTMCs and CTMDPs: a lyapunov approach. In: McIver, A., Horvath, A. (eds.) QEST 2018. LNCS, vol. 11024, pp. 389–406. Springer, Cham (2018). https://doi.org/10.1007/978-3-319-99154-2_24

160. Alexandre dit Sandretto, J., Chapoutot, A.: Validated explicit and implicit Runge-Kutta methods. Reliable Comput. Electron. Edition 22 (2016)

161. Schupp, S., Abraham, E., Ben Makhlouf, I., Kowalewski, S.: HyPro: a C++ library for state set representations for hybrid systems reachability analysis. In: Proceedings of the NASA Formal Methods Symposium, pp. 288–294 (2017)

162. Sheng, H., Bentkamp, A., Zhan, B.: HHLPy: practical verification of hybrid systems using hoare logic. In: Chechik, M., Katoen, J.P., Leucker, M. (eds.) FM 2023. LNCS, vol. 14000, pp. 160–178. Springer, Cham (2023). https://doi.org/10.1007/978-3-031-27481-7_11

163. Shmarov, F., Zuliani, P.: ProbReach: verified probabilistic δ-reachability for stochastic hybrid systems. In: HSCC, pp. 134–139. ACM (2015)

164. Sidrane, C., Kochenderfer, M.J.: OVERT: verification of nonlinear dynamical systems with neural network controllers via overapproximation. In: Safe Machine Learning Workshop at ICLR (2019)

165. Sogokon, A., Mitsch, S., Tan, Y.K., Cordwell, K., Platzer, A.: Pegasus: sound continuous invariant generation. Formal Methods Syst. Des. **58**(1–2), 5–41 (2021). https://doi.org/10.1007/s10703-020-00355-z
166. Soudjani, S., Gevaerts, C., Abate, A.: FAUST^2: formal abstractions of uncountable-STate STochastic processes. In: TACAS, vol. 15, pp. 272–286 (2015)
167. Strauss, M., Mitsch, S.: Slow down, move over: a case study in formal verification, refinement, and testing of the responsibility-sensitive safety model for self-driving cars. In: Tests and Proofs - International Conference (TAP), pp. 149–167 (2023). https://doi.org/10.1007/978-3-031-38828-6_9
168. Tan, Y.K., Mitsch, S., Platzer, A.: Verifying switched system stability with logic. In: ACM International Conference on Hybrid Systems: Computation and Control (HSCC), pp. 2:1–2:11 (2022). https://doi.org/10.1145/3501710.3519541
169. Thibeault, Q., Anderson, J., Chandratre, A., Pedrielli, G., Fainekos, G.: PSY-TaLiRo: a python toolbox for search-based test generation for cyber-physical systems. In: Lluch Lafuente, A., Mavridou, A. (eds.) FMICS 2021. LNCS, vol. 12863, pp. 223–231. Springer, Cham (2021). https://doi.org/10.1007/978-3-030-85248-1_15
170. Tran, H.-D., et al.: NNV: the neural network verification tool for deep neural networks and learning-enabled cyber-physical systems. In: Lahiri, S.K., Wang, C. (eds.) CAV 2020. LNCS, vol. 12224, pp. 3–17. Springer, Cham (2020). https://doi.org/10.1007/978-3-030-53288-8_1
171. Van Huijgevoort, B., Schön, O., Soudjani, S., Haesaert, S.: Syscore: synthesis via stochastic coupling relations. In: International Conference on Hybrid Systems: Computation and Control. HSCC 2023. ACM (2023). https://doi.org/10.1145/3575870.3587123
172. Vinod, A.P., Gleason, J.D., Oishi, M.M.: Sreachtools: a MATLAB stochastic reachability toolbox. In: ACM International Conference on Hybrid Systems: Computation and Control, pp. 33–38 (2019)
173. Waga, M.: Falsification of cyber-physical systems with robustness-guided black-box checking. In: International Conference on Hybrid Systems: Computation and Control (HSCC), pp. 11:1–11:13. ACM (2020).https://doi.org/10.1145/3365365.3382193
174. Wang, S., Zhan, N., Zou, L.: An improved HHL prover: an interactive theorem prover for hybrid systems. In: Butler, M., Conchon, S., Zaïdi, F. (eds.) ICFEM 2015. LNCS, vol. 9407, pp. 382–399. Springer, Cham (2015). https://doi.org/10.1007/978-3-319-25423-4_25
175. Wetzlinger, M., Kochdumper, N., Althoff, M.: Adaptive parameter tuning for reachability analysis of linear systems. In: IEEE Conference on Decision and Control, pp. 5145–5152 (2020). https://doi.org/10.1109/CDC42340.2020.9304431
176. Wetzlinger, M., Kochdumper, N., Bak, S., Althoff, M.: Fully automated verification of linear systems using inner and outer approximations of reachable sets. IEEE Trans. Autom. Control **68**(12), 7771–7786 (2023). https://doi.org/10.1109/TAC.2023.3292008
177. Wetzlinger, M., Kochdumper, N., Bak, S., Althoff, M.: Fully-automated verification of linear systems using reachability analysis with support functions. In: Proceedings of the 26th ACM International Conference on Hybrid Systems: Computation and Control (2023). https://doi.org/10.1145/3575870.3587121
178. Wetzlinger, M., Kulmburg, A., Althoff, M.: Adaptive parameter tuning for reachability analysis of nonlinear systems. In: International Conference on Hybrid Systems: Computation and Control. HSCC 2021. Association for Computing Machinery (2021). https://doi.org/10.1145/3447928.3456643

179. Wetzlinger, M., Kulmburg, A., Le Penven, A., Althoff, M.: Adaptive reachability algorithms for nonlinear systems using abstraction error analysis. Nonlinear Anal. Hybrid Syst. **46** (2022).https://doi.org/10.1016/j.nahs.2022.101252

180. Winter, S., et al.: A retrospective study of one decade of artifact evaluations. In: ESEC/FSE 2022: Proceedings of the 30th ACM Joint European Software Engineering Conference and Symposium on the Foundations of Software Engineering, pp. 145–156. ACM (2022). https://doi.org/10.1145/3540250.3549172

181. Yamagata, Y., Liu, S., Akazaki, T., Duan, Y., Hao, J.: Falsification of cyber-physical systems using deep reinforcement learning. IEEE Trans. Software Eng. **47**(12), 2823–2840 (2021). https://doi.org/10.1109/TSE.2020.2969178

182. Zhang, Z., Lyu, D., Arcaini, P., Ma, L., Hasuo, I., Zhao, J.: Effective hybrid system falsification using Monte Carlo tree search guided by QB-robustness. In: Silva, A., Leino, K.R.M. (eds.) CAV 2021. LNCS, vol. 12759, pp. 595–618. Springer, Cham (2021). https://doi.org/10.1007/978-3-030-81685-8_29

183. zlscheck (2023). https://github.com/ismailbennani/zlscheck

Competition of Solvers for Constrained Horn Clauses (CHC-COMP 2023)

Emanuele De Angelis[1]($\boxtimes$)(iD) and Hari Govind Vediramana Krishnan[2]($\boxtimes$)(iD)

[1] IASI-CNR, Rome, Italy
emanuele.deangelis@iasi.cnr.it
[2] University of Waterloo, Waterloo, Canada
hgvedira@uwaterloo.ca

Abstract. Constrained Horn Clauses (CHC) are a fragment of first-order logic that is expressive enough to represent many important software verification tasks, yet practical for fully automatic techniques. CHCs are used to verify the safety of hardware, software, and hybrid systems, infer refinement types, analyse program termination, and prove the correctness of smart contracts, among other applications. Over the years, many researchers have developed a variety of tools to check the satisfiability of CHCs. CHC-COMP strives to bring together these research efforts and compare the effectiveness of different CHC solvers on a unified set of benchmarks. Each year, CHC-COMP collects benchmarks provided by the CHC community, organizes them into tracks, and evaluates CHC solvers against those tracks. This annual competition is now in its sixth year. This report gives an overview of CHC-COMP and details of its sixth edition.

Keywords: Constrained Horn Clauses · Constraint Logic Programming · Satisfiability Modulo Theories · Constraint Solving

1 Introduction

Constrained Horn Clauses (CHCs) constitute a fragment of the First Order Predicate Calculus, where the Horn clause syntax is extended by allowing *constraints*, that is, formulae whose interpretation is defined in a, possibly non-Horn, *background* theory, which may occur in the premises of clauses. More precisely, a CHC (or simply, a clause) is a universally quantified implication of the form $\forall(\varphi \wedge L \rightarrow H)$. The conclusion (or *head*) H is either an atomic formula (*atom*, for short) or *false*. The premise (or *body*) $\varphi \wedge L$ is a conjunction of a constraint φ, and a (possibly empty) conjunction L of atoms. A clause whose head is *false* is called a *query*. If L consists of at most one atom, the clause is called *linear*. Otherwise, it is called *nonlinear*. By following logic programming notation, a CHC of the form $\forall(\varphi \wedge L \rightarrow H)$ is written as $H \leftarrow \varphi, L$.

CHCs have gained popularity as a formalism well suited for automatic program analysis and verification [3,5]. Indeed, by allowing constraints of an arbitrary background theory (such as linear integer arithmetic, arrays, algebraic

D. Beyer et al. (Eds.): TOOLympics 2024, LNCS 14550, pp. 38–51, 2025.
https://doi.org/10.1007/978-3-031-67695-6_2

data types, and combinations thereof), CHCs turn out to be a very convenient *intermediate language* to effectively perform a variety of program analysis and verification tasks (such as proving correctness and termination [7–9,11,16,18,20,22,23,33,36] as well as performing complexity and resource analysis [15,31]) that can be reduced to the satisfiability problem for CHCs[1]. For instance, the verification conditions for proving the validity of the Hoare triple $\{n \geq 1\}$ x=0; y=0; while(x<n) { x=x+1; y=y+2; } $\{y > x\}$, where x, y and n are integer variables, can be expressed by the following CHCs with constraints interpreted in the theory of linear integer arithmetic (LIA):

$$p(X, Y, N) \leftarrow X = 0 \wedge Y = 0 \wedge N \geq 1$$
$$p(X+1, Y+2, N) \leftarrow X < N, \ p(X, Y, N)$$
$$false \leftarrow X \geq N \wedge Y \leq X, \ p(X, Y, N)$$

Thus, the problem of proving the validity of the above Hoare triple reduces to the problem of proving the *satisfiability* of the corresponding CHCs, that is, finding an interpretation for the uninterpreted predicate p over LIA (that is, a relation specified by constraints) that makes the three clauses *true*. For instance, one such interpretation for $p(X, Y, N)$ is $(X < Y \vee X < N) \wedge X < Y + 1$.

In the last decade, advances in the field have led to the development of several very powerful solvers for CHCs, which can nowadays be effectively used as back-end tools for program verification purposes due to their ability to solve satisfiability problems over a variety of background theories. A non-exhaustive list of tools specialized in solving CHCs includes: ADTInd [38], ADTRem [10], Eldarica [25], FreqHorn [16], Golem [4] HSF [19], PCSat [36], RAHFT [28], RInGen [30], SPACER [29], Ultimate TreeAutomizer [14], and VeriMAP [6].

The *Competition of Solvers for Constrained Horn Clauses* (CHC-COMP[2]) is an annual contest that aims to evaluate the state of the art in tools for solving CHCs. It is open to proposals and contributions from users and developers of CHC solvers, as well as researchers working in the field of CHC solving foundations and its applications.

CHC-COMP is organized in tracks, each of which deals with a class of CHCs. CHCs are classified according to the form of their clauses (either linear or nolinear), and the background theory of their constraints. The performance of the participating solvers are evaluated on a set of track-specific benchmarks, which are selected by the organizers from all available benchmarks. The winners among the competing solvers are determined by counting the number of solved problems and, in the case of ex-aequo, by also considering the solving time.

CHC-COMP 2023 was affiliated with HCVS 2023[3] held in Paris, France, on April 23, 2023. The CHC-COMP 2023 deadline for submitting benchmarks to

---

[1] CHCs are syntactically and semantically the same as *Constraint Logic Programs* (CLP) [26]. The difference in the use of the two terms is purely pragmatic: in the literature, CHCs is often used instead of CLP when the focus is mainly on the logical meaning of clauses encoding verification problems (and, in particular, in the construction of models for the clauses) rather than their execution as programs.

[2] The CHC-COMP webpage is available at https://chc-comp.github.io/.

[3] The HCVS 2023 webpage is available at https://www.sci.unich.it/hcvs23/.

be considered for the competition was March 24, 2023. The deadline for submitting the solvers for test runs (optional) was March 31, 2023. The deadline for submitting the solvers for the competition runs was April 7, 2023. The competition was run in the subsequent two weeks, and the results were announced at HCVS 2023. CHC-COMP 2023 featured 7 solvers (6 competing solvers and one hors concours), and 6 tracks consisting of linear and nonlinear CHCs with constraints over linear integer arithmetic, arrays, algebraic data types, and a few combinations thereof.

Organization of the Paper. The remaining part of this paper is organized as follows. Section 2 describes how the CHC-COMP is organized and run: the competition setup, the scoring and ranking scheme, and the technical resources used to evaluate the solvers. Section 3 presents the competition tracks, and how the candidate benchmarks have been processed and selected for the competition runs. Sections 4 and 5 present the tools that entered CHC-COMP 2023 and its results, respectively. Section 6 concludes the paper with a few notes and remarks.

2 Organization

CHC-COMP is run off-site on controlled resources: solvers are submitted to the competition in the form of StarExec packages, which is the platform used to run the competition and make the results available to the CHC community.

CHC-COMP is affiliated with the Workshop on Horn Clauses for Verification and Synthesis, where the organizers can announce the results and present the report of the competition.

In the remainder of this section, we describe the process adopted for evaluating the solvers as well as the ranking scheme used to select the winners.

2.1 Evaluating Solvers: Test and Competition Runs

CHC solvers are evaluated by performing a *test* run and a *competition* run on the StarExec platform. A run involves submitting jobs to StarExec, that is, collections of ⟨solver-configuration, benchmark⟩ pairs. The data gathered from the 'job information' CSV files produced by StarExec are used to rank the solvers.

The *test* run is used by the participants to get acquainted with the StarExec platform and test out their pre-submissions. Submitting a solver for *test* runs is optional. During this test phase, the organizers contact the participants if they find any issues with their submission so that the participants can fix it before their final submission. The participants are given a week in between the *test* and *competition* runs. In the *test* runs, a small set of randomly selected benchmarks is used, and each job is limited to 600 s CPU time, 600 s wall-clock time, and 64 GB memory.

In *competition* runs, the final submissions of the solvers are evaluated to determine the outcome of the competition, that is, to rank the solvers that entered the competition. In these runs each job is limited to 1800s CPU time, 1800s wall-clock time, and 64 GB memory.

Sometimes, the competition benchmarks expose soundness bugs in solvers. We catch these bugs if two solvers disagree on the satisfiability of a benchmark. At CHC-COMP, we keep things friendly by informing the participants about the inconsistency and giving them the benchmark to reproduce the issue. If we have time, we even give them a chance to fix the issue and resubmit their tool. If not, we disqualify the tool from the track.

StarExec made available to CHC-COMP 2023 a queue, called `chcseq.q`, consisting of 20 nodes equipped with Intel(R) Xeon(R) Gold 6334 CPUs[4]. All 'job information' CSV files of the CHC-COMP 2023 runs are available on the StarExec space `CHC/CHC-COMP/CHC-COMP-23`[5].

2.2   Ranking Scheme

The ranking of solvers in each track is based on the score obtained by the solvers in the competition run for a track. The score is computed on the basis of the result provided by the solver on the benchmarks for that track. The result can be *sat*, *unsat*, or *unknown* (which includes solvers giving up, running out of resources, or crashing), and the score is given by the number of *sat* and *unsat* results. In the case of ex-aequo, the ranking is determined by using the CPU time, which is the total CPU time needed by a solver to produce the results.

3   Benchmarks

In this section, we present the competition tracks, the inventory of benchmarks, and the process for selecting the benchmarks for the competition runs.

3.1   Tracks

CHC-COMP is organized in tracks, each of which deals with a class of CHCs. The classification of the CHCs is based on the following features of the clauses:

(i) the background theory of the constraints, and
(ii) the number of *uninterpreted atoms* (that is, atoms whose predicate symbols do not belong to the background theory) occurring in the bodies of the clauses, thereby classifying them as either *linear* (if all clauses have at most one atom in their bodies) or *nonlinear* (if there is at least one clause with more than one atom in its body). In practice, the difference is that linear CHCs encode transition systems and programs whose functions have been inlined and nonlinear CHCs encode programs with function calls. The tracks were separated because 1) the problems are significantly different and 2) some tools only support linear CHCs.

---

[4] https://www.starexec.org/starexec/public/machine-specs.txt.
[5] https://www.starexec.org/starexec/secure/explore/spaces.jsp?id=538944.

CHC-COMP is open to proposals for new competition tracks as well as for changes to the already existing tracks. For CHC-COMP 2023, we had the following tracks:

1. **LIA-lin**: Linear Integer Arithmetic - linear clauses
2. **LIA-nonlin**: Linear Integer Arithmetic - nonlinear clauses
3. **LIA-lin-Arrays**: Linear Integer Arithmetic & Arrays - linear clauses
4. **LIA-nonlin-Arrays**: Linear Integer Arithmetic & Arrays - nonlinear clauses
5. **LIA-nonlin-Arrays-nonrecADT**: Linear Integer Arithmetic & Arrays & nonrecursive Algebraic Data Types - nonlinear clauses
6. **ADT-LIA-nonlin**: Algebraic Data Types & Linear Integer Arithmetic - nonlinear clauses

Table 1. Summary of benchmarks. (*new*) denotes a set of benchmarks added among the available repositories in CHC-COMP 2023.

Repository	LIA-lin	LIA-nonlin	LIA-lin-Arrays	LIA-nonlin-Arrays	LIA-nonlin-Arrays-nonrecADT	ADT-LIA-nonlin
adtrem (*new*)						247
aeval	54					
aeval-unsafe	54					
chc-comp19			290			
eldarica-misc	136	66				
extra-small-lia	55					
hcai	87	131	39	25		
hopv	48	67				
jayhorn	73	7224				
kind2		736				
ldv-ant-med			10	342		
ldv-arrays			2	546		
llreve	66	57	31			
quic3			43			
rust-horn (*new*)	11	6				56
seahorn	2812	66				
solidity					2174	
sv-comp	2930	1169	73	780		
synth/nay-horn		114				
synth/semgus				4839		
tip-adt-lia (*new*)						320
tricera	405	4				
tricera/adt-arrays					156	
ultimate		8		23		
vmt	803					
total	**7534**	**9648**	**488**	**6555**	**2330**	**623**

ADT-LIA-nonlin was introduced in CHC-COMP 2023, and the remaining tracks were inherited from the previous editions.

In addition to the theories occurring in the above list (Linear Integer Arithmetic, Arrays, nonrecursive/recursive Algebraic Data Types, and combinations thereof), benchmarks in all tracks can also make use of the Bool theory.

Finally, in LIA constraints we allow the syntactic appearance of the function symbols *, *div*, *mod*, and *abs*. If these operations do appear, the benchmark is included/excluded from the set of LIA benchmarks according to the following rules: (i) if the second argument of any *div* and *mod* operation is not a constant term, the benchmark is excluded; (ii) if there is more than one non-constant term in any * operation, the benchmark is excluded; (iii) otherwise, the operations are considered semantically linear and the benchmark is included.

Table 2. The number of benchmarks to select and the number of selected benchmarks from each repository. A small number of benchmarks is available for the LIA-lin-Arrays track; therefore, all of the benchmarks have been selected.

Repository	LIA-lin	LIA-nonlin	LIA-lin-Arrays	LIA-nonlin-Arrays	LIA-nonlin-Arrays-nonrecADT	ADT-LIA-nonlin
adtrem						125/125
aeval	30/30					
aeval-unsafe	30/30					
chc-comp19			290/290			
eldarica-misc	45/25	30/26				
extra-small-lia	30/22					
hcai	45/14	60/20	39/39	15/11		
hopv	30/6	30/16				
jayhorn	30/6	180/180				
kind2		90/52				
ldv-ant-med			10/10	60/60		
ldv-arrays			2/2	90/90		
llreve	30/11	45/18	31/31			
quic3			43/43			
rust-horn						28/18
seahorn	90/90	45/15				
solidity					312/127	
sv-comp	90/38	90/48	73/73	135/135		
synth/nay-horn		60/48				
synth/semgus				135/135		
tip-adt-lia						160/160
tricera/svcomp20	60/60	3/0				
tricera/adt-arrays					156/122	
ultimate		6/5		15/15		
vmt	90/90					
to select/**selected**	600/**422**	639/**428**	488/**488**	450/**446**	468/**249**	313/**303**

All benchmarks used for the competition are selected from repositories in the chc-comp[6] GitHub organization. Anyone is welcome to contribute to this repository by adding new benchmarks. Table 1 summarizes the (unique) benchmarks available in each repository.

The organizers are responsible for picking a subset of all available benchmarks for each year's competition. In the rest of this section, we explain the steps in this selection.

3.2   Processing Benchmarks

As mentioned earlier, anyone can submit benchmarks to CHC-COMP. Benchmarks are submitted as SMT files (extension .smt2) following the SMT-LIB 2.6 format [2]. To ensure that submitted benchmarks are indeed in the Horn fragment, we check the syntactic structure of each benchmark. The grammar for our benchmarks is detailed in our website[7]. To help users submit benchmarks, we have tools to check conformance with our format as well as print a given benchmark in the syntactic fragment [1]. Among other restrictions, our format allows exactly one query clause per benchmark. A typical use case for these tools is to merge all query clauses in a benchmark into one clause.

In processing benchmarks, we also classify them in tracks. To check whether a benchmark belongs to a theory, we check for occurrences of interpreted predicate symbols in the theory. As mentioned earlier, we allow the function symbols $*$, *div*, *mod*, and *abs* to appear in benchmarks in linear arithmetic tracks, and we check their arguments to ensure that they are semantically linear. To check linearity of clauses, we merely count the number of syntactic occurrences of uninterpreted atoms in the premise (also called *tail* in the CHC-COMP grammar) of clauses. All scripts used to classify benchmarks are available in [1].

3.3   Selecting Benchmarks

Once all benchmarks in all repositories are classified into tracks, we select a subset of all available benchmarks for the annual competition. In this section, we give a brief overview of the selection process. For a more detailed description of the process, please refer to the report included in the proceedings of the HCVS workshop [12,13].

Since there are multiple repositories to choose from, we set a maximum limit on the number of benchmarks to select from each repository (the first value reported in Table 2). These numbers are based on our knowledge of the benchmarks in each repository, and aim at selecting a representative subset of them from all repositories. We then use a heuristic that tries to select the specified number of benchmarks by taking into account a score estimating their *hardness*, which is based on whether or not the benchmarks are solved by the best solvers of the previous CHC-COMP edition.

---

[6] https://github.com/chc-comp.
[7] https://chc-comp.github.io/format.html.

In particular, for CHC-COMP 2023, to evaluate benchmarks we executed the winning and the runner-up solvers of CHC-COMP 2022 from the appropriate track, both with a time limit of 30 s. We assigned a score to each benchmark based on the outcomes of these runs: a score of 0 if both solvers solve it, 1 if only the winning solver solves it, 2 if only the runner-up solver solves it, and 3 if both solvers time out. We aim to achieve a balanced distribution of benchmarks categorized by their scores. Therefore, we selected the benchmark by their score as follows: 20% with a score of 0, another 20% with a score of 1, an additional 20% with a score of 2, and the remaining 40% with a score of 3 (indicating that neither solver solved it within the 30-second time limit). The maximum number of benchmarks to select, and the final number of benchmarks selected, from each repository are shown in Table 2.

It is not always possible to follow this procedure completely. For example, in tracks with Algebraic Data Types (LIA-nonlin-Arrays-nonrecADT and ADT-LIA-nonlin), only two solvers participated in CHC-COMP 2022 and the submitted version of one of the solvers did not support the SMT-LIB syntax for the benchmarks submitted to CHC-COMP 2023. Therefore, the evaluation of the benchmarks in LIA-nonlin-Arrays-nonrecADT and ADT-LIA-nonlin tracks has been performed using just one solver (the hors concours solver Spacer). We selected 40% of benchmarks from those solved by Spacer (within the time limit of 30 s), and the remaining 60% from those on which Spacer timed out. As another exception, we selected all benchmarks in the LIA-lin-Arrays track, as only a few benchmarks are available.

The intention of our selection process is that each year we pick benchmarks that are hard for the best solvers in the previous year. Ideally, we would have run the solvers with the same timeout as used in the competition (20 min). However, there are over 7500 benchmarks to pick from and we expect several timeouts irrespective of the time limit. Hence, for practical reasons, we set the timeout to 30 s. Even shorter timeouts were used in past editions.

Table 3. Solvers and configurations used in the competition tracks; an empty entry denotes that the solver did not enter the competition in that track. The configuration names have been taken as is from solver submissions.

Solver	LIA-lin	LIA-nonlin	LIA-lin-Arrays	LIA-nonlin-Arrays	LIA-nonlin-Arrays-nonrecADT	ADT-LIA-nonlin
Eldarica	def	def	def	def	def	def
Golem	lia-lin	lia-nonlin				
LoAT	loat_horn					
Theta	fix	fix	fix	fix		
Ultimate TreeAutomizer	default	default	default	default		
Ultimate Unihorn	default	default	default	default		
Spacer	def	def	ARRAYS	ARRAYS	def	def

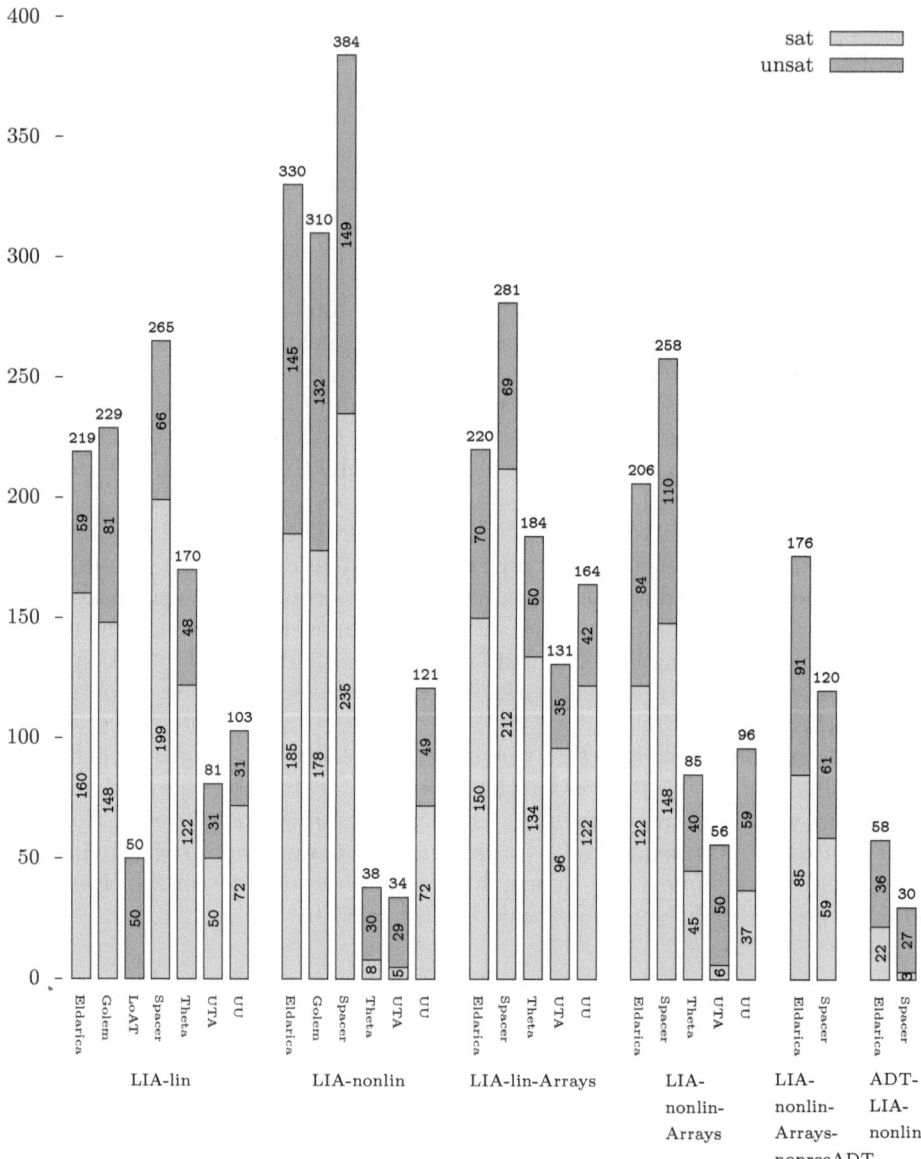

Fig. 1. Results of the CHC-COMP 2023. Each bar represents the result produced by a solver that entered the competition in the track specified below the cluster. The Ultimate Unihorn and Ultimate TreeAutomizer solvers have been abbreviated as UU and UTA, respectively. On the top of the bar it is reported the total number of benchmarks solved: the green segment and the red segment report the number of sat results and unsat results, respectively, produced by a solver. (Color figure online)

Table 4. Results of CHC-COMP 2023.

	LIA-lin	LIA-nonlin	LIA-lin-Arrays	LIA-nonlin-Arrays	LIA-nonlin-Arrays-nonrecADT	ADT-LIA-nonlin
Winner	Golem	Eldarica	Eldarica	Eldarica	Eldarica	Eldarica
2nd place	Eldarica	Golem	Theta	Ultimate Unihorn		
3rd place	Theta	UltimateUnihorn	UltimateUnihorn	Theta		

4   Solvers

Seven CHC solvers were submitted to CHC-COMP 2023: six competing solvers, that is, (1) *Eldarica*[8] [25], (2) *Golem*[9], (3) *LoAT*[10] [17], (4) *Theta*[11] [21,32,35], (5) *Ultimate TreeAutomizer*[12] [14,27,37], (6) *Ultimate Unihorn*[13] [24], and one solver *hors concours*, that is, (7) *Spacer*[14] [29], co-developed by Hari Govind V K who is co-organizing the CHC-COMP 2023 and therefore can only compete unofficially, as usual at CHC-COMP.

Table 3 lists the solvers and, for each CHC-COMP 2023 track, it reports the name of the solver configuration file used to run the solver on that track.

The binaries of the solvers are available on the StarExec space `CHC/CHC-COMP /CHC-COMP-23-competitions-runs`[15].

5   Results

Figure 1 shows the number of (sat and unsat) benchmarks solved by each participating solver in each track. Table 4 summarizes the results of CHC-COMP 2023. All the data gathered from the execution of the StarExec jobs created to run the competition are available on StarExec[16].

The tables included in the presentation of the CHC-COMP 2023 report at HCVS[17] include detailed data about the competition runs. Spacer, which entered the competition as hors concours solver, placed in the first position of the following tracks: LIA-lin, LIA-nonlin, LIA-lin-Arrays, and LIA-nonlin-Arrays.

---

[8] https://github.com/uuverifiers/eldarica.
[9] https://github.com/usi-verification-and-security/golem.
[10] https://loat-developers.github.io/LoAT/.
[11] https://github.com/ftsrg/theta.
[12] https://ultimate.informatik.uni-freiburg.de/.
[13] See link for *Ultimate TreeAutomizer* (see footnote 12).
[14] part of Z3 SMT solver: https://github.com/Z3Prover/z3.
[15] https://www.starexec.org/starexec/secure/explore/spaces.jsp?id=538230.
[16] space `CHC/CHC-COMP/CHC-COMP-23-competitions-runs`.
[17] https://chc-comp.github.io/CHC-COMP2023Report-HCSV.pdf, https://chc-comp.github.io/CHC_COMP_2023_Competition_Report.pdf.

6   Conclusions

The Competition of Solvers for Constrained Horn Clauses (CHC-COMP) is an annual competition aimed at bringing together developers and users of CHC solvers. The sixth edition of the CHC-COMP (CHC-COMP 2023) has seen the participation of seven solvers. Solvers entered six tracks, each of which deals with a different class of CHCs, that is, including either linear or nonlinear clauses with constraints over four background theories: Linear Integer Arithmetic, Arrays, non-recursive Algebraic Data Types, recursive Algebraic Data Types. As in previous years, all benchmarks contributed by the CHC community were processed to make them compliant with the CHC-COMP format. Then, the hardness of benchmarks was evaluated by checking whether or not the winner and the runner-up solvers of the previous edition of CHC-COMP (CHC-COMP 2022) were able to solve them within a timeout of 30 s. Benchmarks for each track were selected by taking into consideration the hardness of the problems as well as the number of benchmarks included in each repository and their variety. All participating solvers were run on selected benchmarks and results announced at HCVS 2023.

The main open issue to address in the future is establishing a method to validate the correctness of the answers produced by the CHC solvers. Several solvers have support for generating a witness (a model or counterexample). However, the witness is used mainly for debugging by the developers. Sometimes, these witnesses are not for the original CHCs but for the CHCs obtained after many layers of pre-processing. Transforming these "internal" witnesses into a witness for the original problem is a work in progress.

Acknowledgements. Since its very first edition in 2018, CHC-COMP has been affiliated with the Workshop on Horn Clauses for Verification and Synthesis (HCVS). Presenting the CHC-COMP report at HCVS allowed the organizers, besides announcing the winners, to receive fruitful feedback from the researchers developing and using CHC solvers. We thank all organizers of HCVS for hosting CHC-COMP.

CHC-COMP is now in its sixth edition. Previous editions of the competition were: CHC-COMP 2018, organized by Arie Gurfinkel (University of Waterloo, Canada), Philipp Rümmer (Uppsala University, Sweden), Grigory Fedyukovich (Florida State University, USA), and Adrien Champion (University of Tokyo, Japan); CHC-COMP 2019, organized by Grigory Fedyukovich; CHC-COMP 2020, organized by Philipp Rümmer; CHC-COMP 2021, organized by Grigory Fedyukovich and Philipp Rümmer; CHC-COMP 2022, organized by Emanuele De Angelis (IASI-CNR, Italy) and Hari Govind V K (University of Waterloo, Canada). CHC-COMP 2023 builds upon the infrastructure developed by the organizers of the previous editions, which also includes the contributions from Nikolaj Bjørner and Dejan Jovanovic. We thank all previous organizers for their continuing support in organizing CHC-COMP.

We would like to thank all the teams that submitted their solvers to CHC-COMP 2023, as well as who contributed with benchmarks. We would also like to thank the HCVS 2023 Program Chairs, David Monniaux and Jose F. Morales, for hosting the competition this year as well, and all the HCVS attendees for the fruitful discussion we had after the presentation of the CHC-COMP report. A special thanks goes to Hossein Hojjat for presenting CHC-COMP 2023 at TOOLympics.

We are also extremely grateful to StarExec (https://www.starexec.org/) [34] that continues to support the CHC community by providing the CHC-COMP the computing resources to run the solvers. In particular, we would like to thank Aaron Stump for helping us in accessing and using the StarExec services.

Emanuele De Angelis is member of the INdAM Research Group GNCS. Hari Govind V K was supported by the Microsoft Research PhD Fellowship.

References

1. chc-tools github repository (2019). https://github.com/chc-comp/chc-tools.git
2. Barrett, C., Fontaine, P., Tinelli, C.: The Satisfiability Modulo Theories Library (SMT-LIB) (2016). www.SMT-LIB.org
3. Bjørner, N., Gurfinkel, A., McMillan, K., Rybalchenko, A.: Horn clause solvers for program verification. In: Beklemishev, L.D., Blass, A., Dershowitz, N., Finkbeiner, B., Schulte, W. (eds.) Fields of Logic and Computation II. LNCS, vol. 9300, pp. 24–51. Springer, Cham (2015). https://doi.org/10.1007/978-3-319-23534-9_2
4. Blicha, M., Britikov, K., Sharygina, N.: The golem horn solver. In: Enea, C., Lal, A. (eds.) CAV 2023. LNCS, vol. 13965, pp. 209–223. Springer, Cham (2023). https://doi.org/10.1007/978-3-031-37703-7_10
5. De Angelis, E., Fioravanti, F., Gallagher, J.P., Hermenegildo, M.V., Pettorossi, A., Proietti, M.: Analysis and transformation of constrained Horn clauses for program verification. Theory Pract. Logic Program. **22**(6), 974–1042 (2022). https://doi.org/10.1017/S1471068421000211
6. De Angelis, E., Fioravanti, F., Pettorossi, A., Proietti, M.: VeriMAP: a tool for verifying programs through transformations. In: Ábrahám, E., Havelund, K. (eds.) TACAS 2014. LNCS, vol. 8413, pp. 568–574. Springer, Heidelberg (2014). https://doi.org/10.1007/978-3-642-54862-8_47
7. De Angelis, E., Fioravanti, F., Pettorossi, A., Proietti, M.: A rule-based verification strategy for array manipulating programs. Fund. Inform. **140**(3–4), 329–355 (2015). https://doi.org/10.3233/FI-2015-1257
8. De Angelis, E., Fioravanti, F., Pettorossi, A., Proietti, M.: Semantics-based generation of verification conditions via program specialization. Sci. Comput. Program. **147**, 78–108 (2017). https://doi.org/10.1016/j.scico.2016.11.002
9. De Angelis, E., Fioravanti, F., Pettorossi, A., Proietti, M.: Predicate Pairing for program verification. Theory Pract. Logic Program. **18**(2), 126–166 (2018). https://doi.org/10.1017/S1471068417000497
10. De Angelis, E., Fioravanti, F., Pettorossi, A., Proietti, M.: Satisfiability of constrained Horn clauses on algebraic data types: a transformation-based approach. J. Log. Comput. **32**(2), 402–442 (2022). https://doi.org/10.1093/logcom/exab090
11. De Angelis, E., Fioravanti, F., Pettorossi, A., Proietti, M.: Verifying catamorphism-based contracts using constrained Horn clauses. Theory Pract. Logic Program. **22**(4), 555–572 (2022). https://doi.org/10.1017/S1471068422000175
12. De Angelis, E., Hari Govind V K: CHC-COMP 2022: competition report. Electron. Proc. Theor. Comput. Sci. **373**, 44–62 (2022). https://doi.org/10.4204/eptcs.373.5
13. De Angelis, E., Hari Govind, V.K.: CHC-COMP 2023: competition report. Electron. Proc. Theor. Comput. Sci. **402**, 83–104 (2024). https://doi.org/10.4204/EPTCS.402.10

14. Dietsch, D., Heizmann, M., Hoenicke, J., Nutz, A., Podelski, A.: Ultimate TreeAutomizer (CHC-COMP tool description). In: De Angelis, E., Fedyukovich, G., Tzevelekos, N., Ulbrich, M. (eds.) Proceedings of the Sixth Workshop on Horn Clauses for Verification and Synthesis and Third Workshop on Program Equivalence and Relational Reasoning, HCVS/PERR@ETAPS 2019, Prague, Czech Republic, 6–7 April 2019. EPTCS, vol. 296, pp. 42–47 (2019). https://doi.org/10.4204/EPTCS.296.7

15. Doménech, J.J., Gallagher, J.P., Genaim, S.: Control-flow refinement by partial evaluation, and its application to termination and cost analysis. Theory Pract. Logic Program. **19**(5–6), 990–1005 (2019). https://doi.org/10.1017/S1471068419000310

16. Fedyukovich, G., Zhang, Y., Gupta, A.: Syntax-guided termination analysis. In: Chockler, H., Weissenbacher, G. (eds.) CAV 2018. LNCS, vol. 10981, pp. 124–143. Springer, Cham (2018). https://doi.org/10.1007/978-3-319-96145-3_7

17. Frohn, F., Giesl, J.: Proving non-termination and lower runtime bounds with LoAT (system description). In: Blanchette, J., Kovács, L., Pattinson, D. (eds.) IJCAR 2022. LNCS, vol. 13965, pp. 712–722. Springer, Cham (2022). https://doi.org/10.1007/978-3-031-10769-6_41

18. Gange, G., Navas, J.A., Schachte, P., Søndergaard, H., Stuckey, P.J.: Horn clauses as an intermediate representation for program analysis and transformation. Theory Pract. Logic Program. **15**(4–5), 526–542 (2015). https://doi.org/10.1017/S1471068415000204

19. Grebenshchikov, S., Lopes, N.P., Popeea, C., Rybalchenko, A.: Synthesizing software verifiers from proof rules. In: Vitek, J., Lin, H., Tip, F. (eds.) ACM SIGPLAN Conference on Programming Language Design and Implementation, PLDI 2012, Beijing, China - 11–16 June 2012, pp. 405–416. ACM (2012). https://doi.org/10.1145/2254064.2254112

20. Gurfinkel, A.: Program verification with constrained Horn clauses (invited paper). In: Shoham, S., Vizel, Y. (eds.) CAV 2022. LNCS, vol. 13371, pp. 19–29. Springer, Cham (2022). https://doi.org/10.1007/978-3-031-13185-1_2

21. Hajdu, Á., Micskei, Z.: Efficient strategies for cegar-based model checking. J. Autom. Reason. **64**(6), 1051–1091 (2020). https://doi.org/10.1007/s10817-019-09535-x

22. Hamza, A., Fedyukovich, G.: Lockstep composition for unbalanced loops. In: Sankaranarayanan, S., Sharygina, N. (eds.) TACAS 2023. LNCS, vol. 13994, pp. 270–288. Springer, Cham (2023). https://doi.org/10.1007/978-3-031-30820-8_18

23. Hari Govind V. K., Shoham, S., Gurfinkel, A.: Solving constrained Horn clauses modulo algebraic data types and recursive functions. Proc. ACM Program. Lang. **6**(POPL), 1–29 (2022). https://doi.org/10.1145/3498722

24. Heizmann, M., et al.: Ultimate automizer and the commuhash normal form - (competition contribution). In: Sankaranarayanan, S., Sharygina, N. (eds.) TACAS 2023, Part II. LNCS, vol. 13994, pp. 577–581. Springer, Cham (2023). https://doi.org/10.1007/978-3-031-30820-8_39

25. Hojjat, H., Rümmer, P.: The ELDARICA Horn solver. In: 2018 Formal Methods in Computer Aided Design, FMCAD, pp. 1–7 (2018). https://doi.org/10.23919/FMCAD.2018.8603013

26. Jaffar, J., Maher, M.J.: Constraint logic programming: a survey. J. Log. Program. **19**(20), 503–581 (1994). https://doi.org/10.1016/0743-1066(94)90033-7

27. Kafle, B., Gallagher, J.P.: Tree automata-based refinement with application to horn clause verification. In: D'Souza, D., Lal, A., Larsen, K.G. (eds.) VMCAI

2015. LNCS, vol. 8931, pp. 209–226. Springer, Heidelberg (2015). https://doi.org/10.1007/978-3-662-46081-8_12

28. Kafle, B., Gallagher, J.P., Morales, J.F.: RAHFT: a tool for verifying horn clauses using abstract interpretation and finite tree automata. In: Chaudhuri, S., Farzan, A. (eds.) CAV 2016. LNCS, vol. 9779, pp. 261–268. Springer, Cham (2016). https://doi.org/10.1007/978-3-319-41528-4_14

29. Komuravelli, A., Gurfinkel, A., Chaki, S.: SMT-based model checking for recursive programs. Formal Methods Syst. Des. **48**(3), 175–205 (2016). https://doi.org/10.1007/s10703-016-0249-4

30. Kostyukov, Y., Mordvinov, D., Fedyukovich, G.: Beyond the elementary representations of program invariants over algebraic data types. In: Proceedings of the 42nd ACM SIGPLAN International Conference on Programming Language Design and Implementation, PLDI 2021, pp. 451–465. ACM (2021). https://doi.org/10.1145/3453483.3454055

31. López-García, P., Darmawan, L., Klemen, M., Liqat, U., Bueno, F., Hermenegildo, M.V.: Interval-based resource usage verification by translation into Horn clauses and an application to energy consumption. Theory Pract. Logic Program. **18**(2), 167–223 (2018). https://doi.org/10.1017/S1471068418000042

32. Somorjai, M., Dobos-Kovács, M., Ádám, Z., Bajczi, L., Vörös, A.: Bottoms up for CHCs: novel transformation of linear constrained Horn clauses to software verification. Electron. Proc. Theor. Comput. Sci. (2023)

33. Spoto, F., Mesnard, F., Payet, É.: A termination analyzer for java bytecode based on path-length. ACM Trans. Program. Lang. Syst. **32**(3), 8:1–8:70 (2010). https://doi.org/10.1145/1709093.1709095

34. Stump, A., Sutcliffe, G., Tinelli, C.: StarExec: a cross-community infrastructure for logic solving. In: Demri, S., Kapur, D., Weidenbach, C. (eds.) IJCAR 2014. LNCS (LNAI), vol. 8562, pp. 367–373. Springer, Cham (2014). https://doi.org/10.1007/978-3-319-08587-6_28

35. Tóth, T., Hajdu, Á., Vörös, A., Micskei, Z., Majzik, I.: Theta: a framework for abstraction refinement-based model checking. In: 2017 Formal Methods in Computer Aided Design (FMCAD), pp. 176–179 (2017). https://doi.org/10.23919/FMCAD.2017.8102257

36. Unno, H., Terauchi, T., Koskinen, E.: Constraint-based relational verification. In: Silva, A., Leino, K.R.M. (eds.) CAV 2021. LNCS, vol. 12759, pp. 742–766. Springer, Cham (2021). https://doi.org/10.1007/978-3-030-81685-8_35

37. Wang, W., Jiao, L.: Trace abstraction refinement for solving Horn clauses. Comput. J. **59**(8), 1236–1251 (2016). https://doi.org/10.1093/comjnl/bxw017

38. Yang, W., Fedyukovich, G., Gupta, A.: Lemma synthesis for automating induction over algebraic data types. In: Schiex, T., de Givry, S. (eds.) CP 2019. LNCS, vol. 11802, pp. 600–617. Springer, Cham (2019). https://doi.org/10.1007/978-3-030-30048-7_35

Behind the Scene of the Model Checking Contest, Analysis of Results from 2018 to 2023

Nicolas Amat[1], Elvio Amparore[2], Bernard Berthomieu[3], Pierre Bouvier[4,5],
Silvano Dal Zilio[3], Francis Hulin-Hubard[6], Peter G. Jensen[7], Loig Jezequel[8],

Fabrice Kordon[6(✉)], Shuo Li[9], Emmanuel Paviot-Adet[6,10], Laure Petrucci[11],
Jiří Srba[7], Yann Thierry-Mieg[6], and Karsten Wolf[12]

[1] IMDEA Software Institute, Madrid, Spain
[2] Università di Torino, Turin, Italy
[3] LAAS-CNRS, Université de Toulouse, CNRS, Toulouse, France
[4] Univ. Grenoble Alpes, Inria, CNRS, Grenoble INP, LIG, Grenoble, France
[5] Kalray S.A, Montbonnot-Saint-Martin, France
[6] Sorbonne Université, CNRS, LIP6, 75005 Paris, France
fabrice.kordon@lip6.fr
[7] Aalborg University, Aalborg, Denmark
[8] Université de Nantes, LS2N, UMR CNRS 6004, 44321 Nantes, France
[9] Tongji University, Shanghai, China
[10] Université Paris-Cité, 75006 Paris, France
[11] Univ. Sorbonne Paris Nord, CNRS, LIPN, 93430 Villetaneuse, France
[12] University of Rostock, Rostock, Germany

Abstract. This paper takes you behind the scenes of the Model Checking Contest (MCC), an annual competition focusing on the behavioral analysis of asynchronous systems using state-space exploration and model checking techniques.

The MCC is part of a thriving group of scientific events dedicated to provide a fair evaluation of formal analysis tools, with the goal to push forward the state of the art and provide insights into the evolution of the involved technologies and approaches. We give details on the organization of the competition and on the ways we manage models and formulas. We also take a look at the evolution of the results over the 2018—2023 period, using the wide variety of data we collect each year, and report on the impact the MCC had on the competing tools.

1 Introduction

The Model Checking Contest (MCC) is an annual competition that benchmarks and compares model checking tools. Since its inception in 2011, the MCC has consistently been held alongside the Petri nets conference. However, for the second time, the 2023 edition is a part of the TOOLympics, held under the umbrella of the European Joint Conferences on Theory and Practice of Software (ETAPS).

Scientific competitions like the MCC play a crucial role in the progress of their respective fields. These competitions improve reproducibility and provide a basis for comparing various techniques. They also encourage the development of mature tools, a trend exemplified by the SAT, SMT, and recently, the SV competitions. The MCC

D. Beyer et al. (Eds.): TOOLympics 2024, LNCS 14550, pp. 52–89, 2025.
https://doi.org/10.1007/978-3-031-67695-6_3

is no exception, pushing tool developers to innovate and optimize their tools for better performance.

The MCC focuses on the model checking of concurrent systems. The concurrency in system semantics is an essential aspect of modern computing, given the prevalence of multicore and distributed CPU architectures. Petri nets, with their well-grounded theory and wealth of structural results, offer an ideal formalism for capturing this concurrency and the associated complexity.

In the MCC, model checking seeks to determine if a system satisfies a given property. This could range from safety conditions or invariants to more complex specifications expressed in Computation Tree Logic (CTL) or Linear Temporal Logic (LTL) [14]. The contest also examines global properties such as deadlock detection or liveness, along with metrics on the state space and place markings bounds. These examinations are not purely academic; they reflect practical questions that users might pose about their systems.

Over its decade-long run, the MCC has established itself as a reputable and mature competition. The accumulated data across these years presents a rich source for tracking the evolution and progression of model checking tools and techniques. This paper dives into this data, presenting insights and analyses that shed light on the state of model checking today.

The goal of this paper is to provide an overview on this event as it was operated in 2023 (after more than a decade of experience), to discuss results of the 2023 edition and to present some observations based on the results for the period 2018–2023.

The paper is structured as follows: Sect. 2 sets the definitions of the terms we use and the categories of the MCC; Sect. 3 presents an in depth analysis of the models used as benchmarks and presents how models are collected; Sect. 4 presents the process used to generate formulas that are used as queries in LTL or CTL examinations; Sect. 5 contains subsections for each tool that participated recently (since 2021), with some insight on their strengths and weaknesses provided by the respective tool authors; Sect. 6 presents the result of the 2023 edition of the MCC; Sect. 7 presents a retrospective analysis of the evolution of the contest and its results since 2018 and Sect. 8 concludes the paper with some perspectives for improving future editions of the MCC.

2   Main Definitions Within the Model Checking Contest

Let us first define the vocabulary and concepts used in the Model Checking Contest and in this paper.[1]

Tools. The primary aim of the Model Checking Contest is the evaluation of various tools. Developers submit their tools to contend for medals across diverse categories. While tools can be submitted in multiple configurations or variants, they are deemed as part of a single "family". Within this family, only the top-performing variant vies for a podium position, i.e., being among the top three. However, scores of other variants and those of "reference tools" (see Sect. 6.1) are shared for informational purposes only.

---

[1] More details can also be found in the 2019 report on the MCC [62].

Models and Model Instances. Every participating tool in a specific MCC edition is evaluated against a consistent benchmark suite. The community updates this benchmark annually, comprising **models** and **model instances**:

– A **model** symbolizes either an academic or industrial problem, modeled using (possibly Colored) Petri nets and provided in the PNML standard format [51].
– Each model may have multiple **model instances** which are variants with differing configurations. These changes typically include scaling the initial marking, changes in the structure and scaling the definition of color domains for colored nets. This lets us assess how tools behave when the model is scaled up in complexity.

Some models are simply Petri nets (abbreviated as PT nets in this paper standing for Place Transitions Nets [71]) with integer arc weights and place markings. Other models are colored nets expressed as Symmetric Petri nets [35].

When models are provided as colored nets, the MCC also provides an equivalent PT net computed using the unfolding technique [61]. Some colored model instances however do not have such an equivalent PT net because the resulting net is too large (see Sect. 3 for details).

Each year, the community contributes with new models that are called "surprise models". Model instances from previous years are kept in the benchmark as "known models".

Examinations. Tools are tasked with processing **examinations** on the **model instances**. An examination encompasses one or more queries that yield either Boolean or numeric outcomes. If a tool is unable to resolve a query or examination, it can respond with "do not compete" or "cannot compute".

For 2023, the following six examinations were:

– **StateSpace** examination, where the tool must generate the state space of the system and provide four metrics on the state space of the Petri net: the number of nodes (states) and edges (transitions) in the marking graph, the bound on the number of tokens in any given place, the bound on the total number of tokens in any given reachable state. This category is the oldest category of the MCC, it was introduced in the very first edition of the MCC in 2011.
– **Global Properties** examination, where the tool must answer by a Boolean whether the model instance satisfies five global properties: can the net deadlock, are some place markings invariant (stable marking), is the net one-safe, is it quasi-live (are all transitions fireable at least once) and is the net live. These global properties are all model specific, and while they could be expressed as a set of reachability or CTL queries (one per place or transition) there exist some more effective strategies that can be applied such as structural reductions. While Reachability Deadlock was introduced in 2013, the four other queries were only introduced in 2020.
– **Upper Bounds** examination, where the tool must provide the maximum number of tokens that can mark a given set of places in any reachable marking. There are 16 such queries per model instance and the queries are generated randomly each year.

- **Reachability** examination, where the tool must provide a Boolean answer to queries that are either an invariant (AGp) or a test for reachability of a given situation (EFp). The predicate p is a Boolean combination of comparisons between markings of places (for Cardinality) or fireability of certain transitions (Fireability). [2] The reachability problem is a core problem in model-checking, and is known to be decidable for Petri nets (despite unbounded places). The Reachability examination was introduced in the very first MCC in 2011. It is an examination that typically has more participants than any other, and is thus a highly contested category.
- **CTL** examination, where the tool must provide a Boolean answer to queries that are expressed in Computation Tree Logic (CTL). Again 16 formulas are cardinality based and 16 are fireability based. The category was introduced in 2013.
- **LTL** examination, where the tool must provide a Boolean answer to queries that are expressed as a Linear-time Temporal Logic (LTL) formula. Again 16 formulas are cardinality based and 16 are fireability based. The category was introduced in 2013.

For the Upper Bounds, Reachability, CTL and LTL examinations a new set of formulas is generated every year, so even if the model is a "known model" the formulas are always unknown to the competitors.

Runs. A "run" means a tool's execution applied to a specific model instance for a designated examination. For instance, computing the 16 Cardinality-based CTL formulas for the model instance named Philosophers-PT-001000 constitutes a run. Every run is allotted a maximum of 1 h, except for the Global Properties, which are given half that time for each of the five queries. These runs are constrained to utilize up to 4 CPU cores and 16 GB of RAM.

To ensure uniform comparison metrics such as execution time and memory usage, all tools for a given model instance/examination combination are run on the same machine, despite parallel evaluations across multiple computers or clusters.

3   The Models in the Benchmark

The models in the MCC benchmark are enriched every year with new models provided by the community in reply to a "Call for Models", but models from previous years are kept in the benchmark. In this section we provide an analysis of the models used in the 2023 edition of the Model Checking Contest and their evolution since 2018.

3.1   Evolution of Models Between 2018 and 2023

The MCC benchmark is composed of both colored nets (COL) and place transition nets (PT). Each model comes with a description that explains its origins and the nature of the problem that it tries to capture.

---

[2] In a given run of a tool, 16 such queries are submitted. So with 16 Cardinality queries and 16 Fireability queries we have 32 reachability queries per model instance in total.

	Models			Instances				
						PT		
Year	PT	COL	Total	COL	Native	From COL	Total	Total
2018	69	20	89	180	614	153	767	947
2019	71	22	93	193	666	159	825	1 018
2020	80	23	103	213	837	179	1 016	1 229
2021	90	24	114	230	985	196	1 181	1 411
2022	103	24	127	230	1 191	196	1 387	1 617
2023	105	27	132	252	1 208	218	1 426	1 678

Fig. 1. Evolution of the number of models and model instances in the MCC benchmark.

Figure 1 shows there were in 2023 a total of 132 models in the benchmark, up from 89 in 2018. From these models, a larger number of instances is produced thanks to their parameters. We reach to a total of 1 678 instances in 2023. While the MCC provides equivalent PT nets for COL specifications, some larger scalable COL instances cannot be unfolded due to the explosion in the size of the resulting PT net. Note that there are some COL model instances without a corresponding PT net (e.g. in 2023, only 218 out of 252 COL instances had an associated "PT from COL" version).

3.2 Analysis of the Benchmark Models in 2023

So far, the 132 models featured by the MCC have been submitted by 60 different authors, from 12 countries. This number of contributors leads to a wide variety of models, whether in terms of their origins or the way they are built. Since the MCC benchmark currently covers a wide scope of models, we provide here a broad classification of the models according to 14 application domains. For each domain, we also give (in brackets) its numbers of colored models, PT models and models designed within the context of industrial projects.

– **Biology and Chemistry** (11 PT – 1 industrial): Angiogenesis, CircadianClock, Diffusion2D, DNAWalker, EGFr, ERK, GPPP, MAPK, MAPKbis, PaceMaker, PhaseVariation, ViralEpidemic.
– **Business Process and Automation** (3 COL, 15 PT – 7 industrial): BugTracking, BusinessProcesses, CryptoMiner, FamilyReunion, FMS, HealthRecord, HospitalTriage, HouseConstruction, IBM (4 models), Kanban, Medical, ProductionCell, ParamProductionCell, RobotManipulation, UtilityControlRoom.
– **Distributed Memory and Related Algorithms** (2 COL, 6 PT – 1 industrial): CANConstruction, CANInsertWithFailure, LeafsetExtension, MultiCrashLeafsetExtension, QuasiCertifProtocol, SatelliteMemory, SharedMemory, StigmergyCommit.
– **Elections or Consensus** (1 COL, 4 PT – 2 industrial): Election2020, HirschbergSinclair, NeoElection, Raft, StigmergyElection.
– **Games** (1 COL, 4 PT): DLCRound, DLCShifumi, NQueens, Solitaire, Sudoku.
– **Hardware** (2 COL, 8 PT – 5 industrial): ARMCacheCoherence, ASLink, DiscoveryGPU, GPUForwardProgress, NoC3x3, Ring, SafeBus, TokenRing, UtahNoC, Vasy2003.

- **Operating Systems or Middleware** (2 COL, 2 PT): PolyORBLF, PolyORBNT, SimpleLoadBalancer, SmallOperatingSystem.
- **IoT, Cloud, Reconfiguration** (5 PT – 3 industrial): CloudDeployment, CloudOps-Management, CloudReconfiguration, Planning, SmartHome.
- **Mutual Exclusion** (6 COL, 10 PT): Anderson, DatabaseWithMutex, Dekker, DoubleLock, EisenbergMcGuire, FunctionPointer, GlobalResAllocation, LamportFast-MutEx, Peterson, Philosophers, PhilosophersDyn, ResAllocation, RwMutex, SwimmingPool, Szymanski, TwoPhaseLocking.
- **Network Protocols** (3 COL, 9 PT): CSRepetitions, Echo, HexagonalGrid, HypercubeGrid, HypertorusGrid, IOTPpurchase, NeighborGrid, PermAdmissibility, SquareGrid, TCPcondis, TriangularGrid, VehicularWifi.
- **Security** (7 PT – 7 industrial): DES, ShieldIIPs, ShieldIIPt, ShieldPPPs, ShieldPPPt, ShieldRVs, ShieldRVt.
- **Synchronisations and Message Passing** (9 PT): ClientsAndServers, DBSingle-ClientW, DLCflexbar, FlexibleBarrier, MultiwaySync, RingSingleMessageInMbox, SemanticWebServices, ServersAndClients, SieveSingleMsgMbox.
- **Academic and Synthetic Models** (4 COL, 9 PT): DoubleExponent, Eratosthenes, DrinkVendingMachine, JoinFreeModules, Murphy, PGCD, Referendum, RefineWMG, RERS (4 models).
- **Transportation Systems** (3 COL, 6 PT – 3 industrial) AirplaneLD, AutoFlight, AutonomousCar, BART, BridgeAndVehicles, CircularTrains, EnergyBus, Parking, Railroad.

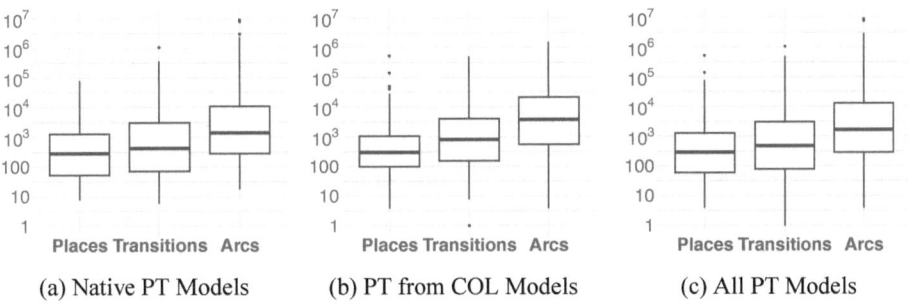

(a) Native PT Models   (b) PT from COL Models   (c) All PT Models

Fig. 2. Log scale box plots of Places, Transitions, and Arcs for PT, unfolded PT from colored models, and all PT nets (including unfolded). The central line of the box represents the median, while the box itself spans the interquartile range (IQR) from the first to the third quartile, covering the middle 50% of the values. Whiskers stretch out to 1.5 times the IQR. Any values exceeding the whiskers are considered outliers and depicted as individual points.

In this section, we delve into the core characteristics of PT model instances in the benchmark, visualized through box plots. Figure 2 showcases the distribution for three principal metrics in the MCC 2023 models: the count of arcs, places, and transitions. Given the significant variance in model sizes, a logarithmic scale was adopted.

For clarity, we have excluded COL model sizes from this visualization, as their pre-unfolding sizes are not directly comparable to PT net dimensions (e.g. whiskers range up to 100 places or transitions at most).

We separate the PT models into "native PT" (Fig. 2a) that were provided in this format, and "PT from COL" (Fig. 2b) that are produced from the COL models up to a certain size. The rightmost plot Fig. 2c merges both of these subcategories. While these plots show that on average models unfolded from COL are larger than native PT nets (in number of arcs), their structure can be expected to be more symmetric, and the largest PT instances in the benchmark are native PT.

The larger PT models in the MCC thus contain up to 10^4 places (with some outliers reaching 10^5 or more), up to 10^5 transitions (with some outliers reaching 10^6), and up to a few million arcs. These plots illustrate the wide variability in the structure size and complexity of PT models.

3.3   A Study of PT Model Instances

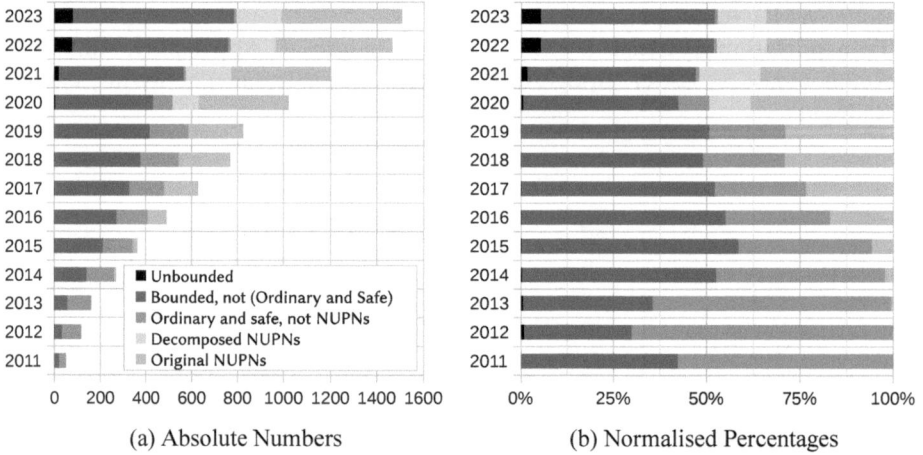

(a) Absolute Numbers   (b) Normalised Percentages

Fig. 3. Distribution of the types of instances provided for each edition of the competition.

Figure 3 offers a yearly breakdown of all PT instances since the MCC's inception, into five mutually exclusives categories. From 2011 to 2023, the first two categories, presented below, accounted for 42% to 58% of PT instances, stabilizing around 50% after 2014:

– Unbounded nets (Black): These nets have an infinite number of reachable markings because of unbounded places, where tokens can accumulate indefinitely. Though they were exceedingly rare before 2020, they now represent 5.4% of all PT instances.

– Bounded, but non-ordinary or non-safe nets (Blue): These are bounded nets that are not ordinary, meaning arc weights can exceed 1, or that are not one-safe, meaning there is at least one reachable marking having several tokens in the same place[3].

The remaining instances are ordinary and one-safe nets, which means they can be expressed as NUPNs (for "Nested-Unit Petri Nets" [45]), an extension of Petri Nets bringing them modularity and hierarchy through a structure of sequential processes (called units), nested in the form of a tree, representing their parent-child relationships. MCC instances are provided in the PNML file format, which enables this NUPN information to be provided through an additional "tool specific" section[4]. These instances are divided into the following three categories:

– Original NUPNs (Green): These are NUPNs stemming from models written in high-level specification languages, which feature the concept of concurrent processes. These processes have been directly translated into NUPN units. In 2023, they constitute 72% of the NUPN instances.
– Decomposed NUPNs (Yellow): These models had their NUPN structures inferred and appended retrospectively, using the 22 decomposition approaches of nets into networks of automata [31], which internally leverage SAT solvers, SMT solvers, and tools for graph coloring and finding maximum cliques[5]. To determine, for a given Petri Net, which NUPN structures are possible, these decomposition approaches depend on the efficient computation of the concurrent place relation [29]. Since 2021, a striking 95.2% of the ordinary and safe instances, lacking an initial NUPN structure, have been upgraded in this manner.
– Non-NUPNs (Red): These nets are ordinary and safe, but they are not described using NUPNs. As of 2021, only ten instances of this category remain, owing to various reasons[6] such as the presence of only trivial decomposition possibilities[6], or their extensive sizes (at present, no participating tool can handle such instances).

4 Formulas in the Benchmark

4.1 The Need and Purpose for Formula Generation

The MCC faces a dichotomy. Whereas models are handcrafted and reused from one year to the next, formulas are randomly generated and changed for each edition of the competition.

In model checking, formulas define the properties of systems, setting clear criteria on the behaviors and conditions a system must meet or avoid. However, even though the

---

[3] It should be noted that some instances are not ordinary, yet are safe, because their initial markings are safe, and all their non-ordinary arcs are connected to dead transitions.

[4] See https://mcc.lip6.fr/nupn.php and https://cadp.inria.fr/man/nupn.html for further details about NUPN file formats.

[5] Notice that subproblems invoked by these decompositions have been used to provide benchmark for the Model Counting Competition [28,49], for the SAT Competition [30,50], and for the SMT-Comp [22,27].

[6] Which provide no additional information to their underlying Petri nets, see [45, Prop. 11].

model forms allow for the declaration of behavioral properties, very few models come with any predefined formulas, and none of the models come with formulas covering all the examinations of the MCC. This poses a challenge that was solved by relying on an automated process for generating new formulas, under the supervision of a dedicated "formula board".

There are some benefits to this situation. Although custom formulas might better align with the model's purpose, a static set of such formulas risks tool over-fitting for known properties. This regular update ensures model checking tools face novel challenges, fostering innovation and avoiding tool stagnation. Notably, the MCC requires 32 formulas for each model instance that span across reachability, CTL, and LTL examinations. This led to the production of 90 912 formulas in 2018, and this number increased to 161 088 by 2023.

Next, we describe the process used for the generation of formulas for the Reachability, CTL, and LTL examinations. While the Upper Bounds examination also utilizes formulas, its generation approach is direct, primarily consisting of selecting places from a net without repetition.

Generation of Reachability and CTL Formulas Using Citili. Reachability and CTL formulas are generated using the tool Citili[7] since 2020. For a given model, Citili proceeds as follows in order to generate one formula.

1. Generate a generic CTL (or Reachability) formula with abstract atoms by randomly selecting operators from a set of allowed operators, up to a given depth.
2. Verify that the formula is acceptable by analyzing its syntax, otherwise generate another one:
    - ensure that it is not in another class of formulas (e.g. it has tree operators for a CTL formula)
    - perform limited checks for triviality of the formula (e.g. tautologies)
3. Instantiate each abstract atom with a concrete one corresponding to the type of formula we want (Cardinality or Fireability).

Building on this, Citili employs the subsequent methodology to generate a challenging formula set of a predefined size (currently set at 16):

1. Generate an initial batch of formulas (currently 32).
2. Set up a rudimentary model checker for each formula:
    - Limit exploration to a set number of states (presently 2 000).
    - Keep formulas that do not produce a result and classify them as challenging.
3. If the stipulated time has not been exhausted and the challenging formula set is below the target size (currently 16), the process returns to step 1.
4. When the time threshold is reached, Citili generates additional formulas to populate the set of challenging formulas. This additional set may include trivial formulas, since they are not subject to any filtering.

---

[7] Available from https://github.com/mcc-petrinets/citili.

Generation of LTL Formulas with the Aid of Spot. The inclusion of a diverse set of LTL formulas is essential to provide a thorough, representative benchmark. In their seminal work, Manna and Pnueli introduced a categorization of LTL formulas, identifying six distinct categories: Reactivity, Recurrence, Persistence, Obligation, Safety, and Guarantee [66]. A balanced representation of these categories ensures that overfitting is avoided and provides a broad coverage of the temporal behaviors in the examination.

Since 2019, we use the following two tools from the SPOT platform [43], in the MCC, to achieve balanced and non-trivial formula generation:

- **ltlrand**: This tool generates random LTL formulas using a predefined set of LTL operators and a designated formula depth. Notably, ltlrand ensures the generated formulas are not trivially reducible, enhancing the complexity of the evaluation process.
- **ltlfilt**: Once the formulas are procured using ltlrand, ltlfilt categorizes each formula based on Manna and Pnueli's classification. This classification step ensures the formulas offer a well-rounded representation across all LTL categories.

Currently, there is no mechanism akin to a "trivial model checker" to evaluate the generated LTL formulas on the first few states of a model instance. However, such an addition can be considered for future improvements, potentially enhancing the quality of formula generation further.

4.2 Analysis of Formulas in 2023

We study in this section a classification of the properties of the MCC into those that can be disproved by exhibiting a counter-example (CEX), and those that must be proved to hold over all reachable configurations or paths (INV). This classification reflects the one for SAT/SMT competitions [50] into SAT (and a model is provided, corresponding to CEX) or UNSAT (which corresponds to INV in our classification).

We study this classification for Global Properties, Reachability and LTL properties. These notions are not applicable to the other categories (CTL, Upper Bounds, State Space). We can only classify a property as being INV or CEX if there is a verdict of at least one tool, so properties no tool could answer are left as Unknown (UNK).

- For "OneSafe" CEX corresponds to existence of a place whose marking can exceed 1,
- for "ReachabilityDeadlock" CEX corresponds to existence of a deadlocked state,
- for "StableMarking" CEX corresponds to existence of a place whose marking never varies,
- for "QuasiLiveness" CEX corresponds to existence of a transition that can never be fired,
- for "Liveness" CEX corresponds to existence of a transition that is not live.
- For "Reachability" properties, an "AG (p)" formula (an invariant that must be true of all states) is a CEX if it is false, and an "EF (p)" formula (a test to see if we can reach a state satisfying p) is a CEX if it is true.
- For "LTL" properties, a false property is a CEX (and we can exhibit a run that does not satisfy the property) and a true property is an INV.

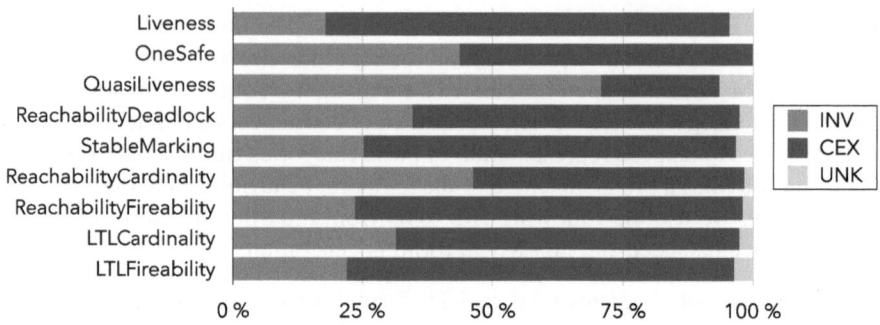

Fig. 4. Distribution of INV vs CEX properties across the different examinations; UNK corresponds to the situation where no tool answered (thus, classification is unknown).

Figure 4 presents the distribution of INV vs CEX for the relevant examinations.

For global properties, the result is purely model-dependent, and in some cases, even instance-dependent. It shows that 56% of the models are one-safe, and that 63% of models contain a deadlock. The models in majority (71%) contain stable places, but most of them (70%) do not contain dead transitions. Only 17% of the models are known to be live, probably due to the abundance of workflow like nets with start and end activities, and models featuring some non repeatable initialization step.

For both Reachability and LTL, there is a clear bias overall for CEX, i.e. properties that can be disproved with a single trace or state. The bias towards CEX is much stronger on Fireability queries (74% of CEX in both reachability and LTL) than on Cardinality queries (52% of CEX for reachability and 65% for LTL). The distribution does evolve from year to year since the properties are generated, but only by a few percent so that qualitatively these observations hold since 2018.

The overall bias towards CEX might skew the results of MCC in favor of tools efficient at bug finding (e.g. directed walks) with respect to strategies that are better at *proving* the system correct (INV), although this is typically a harder problem to solve.

5   Participating Tools

This section presents in more detail the tools that have participated in the MCC since 2021: LoLA, SMPT, TINA.tedd, Enpac, ITS-Tools, TAPAAL, GreatSPN.

5.1   EnPAC

Overview and Evolution. EnPAC (Enhanced Petri-net Analyser and Checker) is a model-checking tool for large concurrent systems modeled as PT nets or their colored extension (COL models). It can evaluate arbitrary queries specified in linear temporal logic (LTL). We started at the end of 2018 and developed an initial version to participate in MCC'2019, which initially supported PT nets. From 2019 to 2020, we extended its capability to colored Petri nets. We directly analyze colored nets without unfolding them to PT nets, which is different from other tools. And we investigated several existing algorithms to improve efficiency, of which the on-the-fly method [46] makes huge progress. From 2020 and 2021, we added encoding strategies (one-safe net encoding, NUPN encoding [45], and P-invariant encoding [85]), and we developed bitwise operations to read and write encoded states, which made encoding strategies more efficient. Additionally, we added heuristic information into Büchi automata so that the counterexample search always follows the shortest path to an acceptable state. From 2021 to 2022, we developed a method where fireable transitions are no longer stored in each state but dynamically generated each time to save memory further.

MCC Impact. The biggest benefit of participating in MCC is that it allows us to know what our weaknesses are. For example, from the result of MCC'2020, we found that EnPAC generally computed faster than other tools but consumed much more memory. Therefore, in the following year, we focused on encoding strategies. We also investigated how to read and write an encoded state using only bit operations, which not only improved memory usage but also reduced the time penalty due to encoding. At present, we recognized the advantages of parallelism shown in MCC and are paying more attention to the sequential-to-parallel bottleneck to extend EnPAC to be parallel.

Availability and Contributors

- **Homepage URL:**
    1. https://github.com/Tj-Cong/EnPAC_2021 (for PT nets)
    2. https://github.com/Tj-Cong/EnPAC_CPN_2021 (for colored nets).
- **License Type:** MIT License.
- **Affiliation:** Tongji University, Shanghai, China.
- **Tool Authors:** Zhijun Ding, Cong He, Shuo Li.
- **Relevant Publications and Contributions:** [48]

5.2   GreatSPN

Overview and Evolution. GreatSPN is an open source framework for Petri nets modeling and analysis. It has several tools for drawing, computing, and verifying different types of Petri nets, either through a graphical interface or through commands. One of the tools is *starMC* [11], a symbolic model checker that uses Multivalued Decision Diagrams (MDD) to generate the state space and check CTL/LTL/CTL* properties. The MDD data structure employed is developed separately in the Meddly library[8], developed at Iowa State university. GreatSPN supports several Petri net formats, including PT nets, generalized stochastic Petri nets, and colored Petri nets.

---

[8] https://github.com/asminer/meddly.

MCC Impact. The *starMC* model checker of GreatSPN has been adapted to compete in all categories of the MCC, albeit it does not have optimized implementations for all of them. The competition for which GreatSPN has the most advanced solution is *StateSpace* generation. The efficient generation of the symbolic reachability graph in GreatSPN is a combination of different factors:

- The representation adopts MDDs to encode the state space;
- The tool uses a specialized *saturation* algorithm with implicit transition relation firings to list all the reachable states;
- An effective heuristic for static variables reordering of the Petri net places in the MDD encoding is used; This heuristic is based on linear algebra properties [9] of the Petri net incidence matrix, and has proven to be highly general and scalable.

GreatSPN also performs model checking of CTL and CTL* properties. LTL expressions can also be checked as CTL* expression. This choice is however suboptimal, since it encounters a significant reduction in the performances.

Availability and Contributors

- **Homepage URL:** https://github.com/greatspn/SOURCES
- **License Type:** GPLv2.
- **Affiliation:** University of Torino, Italy.
- **Tool Authors:** Initially started by Giovanni Chiola, the framework has seen contributions from several developers over its long history. The current main contributors and maintainers are Elvio G. Amparore and Marco Beccuti.
- **Relevant Publications and Contributions:** See [8] for a recent description of the stochastic functionalities of the tool, and [11] for the symbolic model checking algorithms.

5.3   ITS-Tools

Overview and Evolution. ITS-Tools [77] is a model-checker using a portfolio of diverse strategies that include symbolic methods based on hierarchical Set Decision Diagrams [81], constraint based reasoning using the SMT solver Z3 [69] to over-approximate the state space, fast pseudo-random walks to under-approximate it, partial order reduction leveraging the tool LTSMin [59,63], as well as advanced structural reduction rules to reduce the size of the system [79].

The engine has specific support for colored nets using skeleton over-approximations [83] where possible as well as symmetry aware unfoldings that can bypass the model size explosion due to large domains. The LTL engine benefits from several strategies unique to ITS-Tools such as length-sensitivity analysis [70]. For Global Properties and Upper Bounds ITS-Tools benefits from a dedicated set of strategies [80].

MCC Impact. ITS-Tools currently participates in all categories of the MCC. It has participated since the very first edition in 2011 and almost continuously since. While until 2019 Petri net support was offered through a translation to Instantiable Transition Systems (ITS) [77] that give their name to the tool, dedicated Petri net support was gradually introduced to better compete in the MCC.

Availability and Contributors

- **Homepage URL:** https://github.com/lip6/ITSTools
- **License Type:** GPL, EPL
- **Affiliation:** LIP6, Sorbonne Université, CNRS
- **Tool Authors:** Yann Thierry-Mieg with other contributors
- **Relevant Publications and Contributions:** [70,77,79,80]

5.4   LoLA

Overview and Evolution. LoLA (a **L**ow **L**evel Petri net **A**nalyzer) is a model checker for place/transition Petri nets. Its was first released in 1998. LoLA can analyze the state space using stubborn set [74,82], symmetry [75], coverability graph [73], and sweep-line [36,76] reductions. Many formula classes are evaluated by tailored search routines and specific versions of reduction techniques [64,65]. Elements of Petri net theory are used for speeding up verification [84]. Before actual verification, both net and query are simplified using net reduction and abstraction methods [83], as well as Petri net structure theory. LoLA can run several alternative techniques on a query, organized by a portfolio manager [89].

LoLA started as a reachability checker that was soon extended with a CTL model checker. Later, an LTL model checker was added. Then, specific routines for many classes of CTL formulas followed. Most recently, a reduction engine, the portfolio manager, and a skeleton abstraction method became part of LoLA.

MCC Impact. Before the MCC was launched, development of LoLA was mostly driven by our own research. LoLA was the vehicle to generate the "experimental results" table on the second but last page of every paper. LoLA was used in real applications as long as they saw the available techniques fit for their purpose. However, there was no substantial impact of applications to the performance of LoLA.

Through the MCC, we had the opportunity to earn scientific reputation for implementing methods invented by somebody else. This way, the tool grew much more mature and complete. The very competitive nature of the contest pushed us into implementing all the little tricks, and doing all the little optimizations that would have been difficult to publish in a purely theoretical paper. Being quite successful in the MCC, LoLA received a lot of additional attention. The MCC generates very productive exchange of thoughts between tool developers.

Availability and Contributors

- **Homepage URL:** https://theo.informatik.uni-rostock.de/theo-forschung/tools/lola/
- **License Type:** GPL
- **Affiliation:** Universität Rostock, Institut für Informatik, Germany
- **Tool Authors:** Main author is Karsten Wolf. Contributors are mentioned in the source code.
- **Relevant Publications and Contributions:** [85–88]

5.5 SMPT

Overview and Evolution. SMPT (for Satisfiability Modulo Petri Net) is a model checker [4] that participates in the Reachability examination of the MCC since 2021. The tool started as a portfolio of methods to experiment with symbolic, SMT-based model checking techniques, and was designed to be easily extended. Some of its distinctive features are: an adaptation of the Property Directed Reachability (PDR) method for PT nets [5]; and its ability to generate verdict certificates for invariants. It also integrates an approach combining integer linear systems and structural reductions, like in TINA.tedd [16], but adapted to the verification of reachability properties. We give a formal definition of this approach in [2,7], where it is called *polyhedral reduction*, and show how it can be applied to different symbolic methods (BMC, k-induction and PDR), and to more general problems, such as computing the concurrency relation of safe nets [3,6].

At its core, SMPT acts as a front-end to SMT solvers. Since 2022, we added several structural methods, such as invariant checking based on the so-called "state equation method" or with the addition of extra constraints during the verification process, based on results from Petri net theory: structural invariants; traps; invariants originating from the NUPN specification; etc. We also started experimenting with methods based on random walks, which relies on a simulation tool called walk, part of the TINA toolbox [18]. SMPT was again improved in 2023 with the addition of a dedicated method [1] able to transform, in most cases, an initial query (a pair made of a model instance and a reachability formula) into an equivalent query on a reduced, simplified version of the model. It is also the first edition of the MCC where we experimented with formula simplification methods.

MCC Impact. Our participation in the MCC had the effect of transforming what was supposed to be a simple prototype, used for experiments, into a standalone verification tool. It helped us better automate the use of our tool, adding options that simplify the use of SMPT by non-experts and that simplify building strategies to use our portfolio of methods more effectively. It also helped increase the interoperability and the reliability of SMPT, leading us to support use cases that we did not originally envision. For instance supporting models with markings that cannot be represented using 32 bits integers. Finally, it motivated us to expand the perimeter of our tool, and to consider the use of methods outside SMT-based approaches, like random walks, which have been used very effectively by competing tools when checking CEX formulas. By being able to quickly identify classes of queries that are solved more efficiently with other methods, we are able to better focus our efforts on what is, we believe, the strong point of SMPT; the verification of "difficult" invariants (that we could roughly define as INV formulas that are not implied by the state equation).

Availability and Contributors

- **Homepage URL:** https://github.com/nicolasAmat/SMPT
- **License Type:** GNU GPL v3.0
- **Affiliation:** LAAS-CNRS

- **Tool Authors:** Nicolas Amat
- **Relevant Publications and Contributions:** [2, 4]

5.6 TAPAAL

Overview and Evolution. The model checker TAPAAL verifies **T**imed-**A**rc **P**etri nets and is developed at **AAL**borg University in Denmark. The tool was first released in 2008 (see [34, 41]), where it provided a graphical interface for modelling and simulating of timed-arc Petri nets as well as a translation to the UPPAAL-style timed automata while using UPPAAL as the backend engine [33, 52]. In 2011, TAPAAL released its standalone continuous-time engine [42] for reachability and liveness properties. A discrete-time version of this engine [12, 56] was released in 2012. Finally, in 2014 TAPAAL released its own dedicated untimed engine [53] with which it started to participate in the Model Checking Contest.

Initially, TAPAAL participated in the reachability/deadlock category and in 2016 it entered the CTL category with its own on-the-fly CTL model checking algorithm based on dependency graphs [40]. A year later, TAPAAL participated also in the upper-bounds category and finally in 2021 it extended its model checking capabilities to LTL [58]. Nowadays, TAPAAL competes in all categories except state-space size analysis. TAPAAL also supports Petri games [24, 26, 54] and a new release (in preparation) will allow the user to model and verify timed-arc colored Petri games. All verification engines are supported by a GUI that enables us to model, simulate and verify the different extensions of Petri nets, import and export nets in PNML format, create automatic graphical layout, analyze timed workflow nets [67] and many other features.

MCC Impact. TAPAAL's participation in the MCC initiated a significant programming and theoretical research. We have developed a specialised unfolding frontend for our engine in order to deal with colored nets and invented a number of optimization techniques to speedup the unfolding process [19, 20]. In order to store the large state space of reachable markings, we designed a novel data structure PTrie [55] that provides a good compromize between the space efficiency while allowing for fast access times. To efficiently deal with large, randomly generated queries, we carry on an extensive query simplification algorithm [23] as a preprocessing step. This step allows us to simplify, using linear programming, verification queries by identifying subformulae that can never/always be satisfied. Perhaps the most beneficial technique is based on structural reductions [25] that are being continuously improved throughout TAPAAL development. Similarly, new partial order reduction techniques, both for the reachability queries as well as LTL [25, 58], showed a significant improvement. Finally, the MCC benchmark helped us to develop competitive heuristic search strategies [44, 53].

Availability and Contributors

- **Homepage URL:** http://www.tapaal.net
- **License Type:** The GUI is licensed under Open Source Licence 3.0, reduction to timed automata and discrete verification engine is licenced under BSD and

continuous-time and untimed engines are licenced under GPL version 2 and 3, respectively.

- **Affiliation:** Department of Computer Science, Aalborg University, Denmark.
- **Tool Authors:** The development is supervised by Peter G. Jensen, Kenneth Y. Jørgensen and Jiří Srba. A complete list of contributors can be obtained at: https://www.tapaal.net/about/.
- **Relevant Publications and Contributions:** [19,20,23,25,38,39,44,55,57,58]

5.7   TINA.tedd

Overview and Evolution. Tedd is a symbolic model-checker part of the TINA toolbox (TIme Petri Net Analyzer) [17], a set of analysis tools supporting various extensions of Petri nets and Time Petri nets developed at LAAS-CNRS since 1981. TINA provides a wide range of tools for state space generation, structural analysis, model checking, or simulation. It only competes in the StateSpace competition at the MCC.

Tedd plays a role similar to other state-exploration tools provided in TINA, called sift and tina, but based on the use of decision diagrams instead of explicit methods. It is officially part of the TINA release since version 3.7.0 and relies on a dedicated implementation of hierarchical Set Decision Diagrams [81] developed together with Alexandre Hamez. While it only provides support for a limited class of reachability properties, such as finding dead states and transitions, it implements new methods based on the combined use of integer linear systems and structural reductions [15,16] for efficiently counting the number of reachable states and the maximal number of tokens in the marking of places. We credit our uninterrupted first place in the StateSpace examination since 2019 to this new equational technique.

The most important change in the last few years is the move from a sequential portfolio (until 2020) to a parallel portfolio, where we combine tedd with the sift and tina tools at the beginning of each run. Explicit methods can sometimes be faster when there are a few markings, and we are not able to find a good variable order with our symbolic approach. They are also useful for detecting unbounded models. In particular, tina is able to identify all the unbounded model instances found in the current benchmark.

MCC Impact. Beyond enhancing the interoperability and the reliability of our tools, the MCC had a crucial influence on the design and the enhancement of our state space generation technique based on structural reductions, that was first experimented in the 2018 edition of the contest. Since then, we have used new models in this benchmark to experiment with possible reduction rules. More generally, the set of models provided in the MCC have become an invaluable benchmark for testing our tools and comparing new techniques with existing approaches. The MCC benchmark offers many qualities in this respect, because of its impartiality, its representativeness for a large class of use cases, and its scalability (since many models are parameterized). Our participation to the MCC also motivated us to add several tools to our public release, such as an open-source unfolder for colored models [37], and a new dedicated tool, called reduce, that implements the reduction system presented in [16].

Availability and Contributors

- **Homepage URL:** https://projects.laas.fr/tina/
- **License Type:** TINA is freeware; it is closed source, but binary distributions may be freely installed and used.
- **Affiliation:** LAAS-CNRS
- **Tool Authors:** Bernard Berthomieu with other contributors
- **Relevant Publications and Contributions:** [17]

6   Results of the MCC in 2023

6.1   Reference Tools

For the first time during the 2023 edition, the MCC introduced the concept of "Reference Tools". It means tools that are not part of the main competition, and therefore cannot obtain medals, but which are scored and evaluated on the same benchmark. Reference tools may be submitted without the requirement that the submitter be an author of the tool. To test this new opportunity, Y. Thierry-Mieg developed a driver to retro-fit some tools that have competed in previous editions of the MCC. The Reference Tools submitted in 2023 are Marcie, PNMC, LoLA, Smart and LTSMin. All of them have participated before in the MCC but were not submitted as competitors in 2023.

To ensure the fairest possible comparison, the latest stable release of each tool is paired with a driver[9] that provides an unfolding for colored models using ITS-Tools if the back-end tool does not support them natively. The driver also includes a reduction mode, where the model is first processed by ITS-Tools using the strategy described in [78] that uses an SMT solver, some random or directed walks and structural reductions to produce a simpler Petri net and/or formula. The reference tools benefiting from this preprocessing step are indicated with the "+red" suffix.

6.2   Scoring of Tools in 2023

Participating Tools. All together there were fifteen participating tools. Fourteen coming from submissions and one synthetic tool (see below), made from the winners of the previous competition.

- Submitted tools: GreatSPN, ITS-Tools, SMPT, TAPAAL, tedd.
- Reference tools: Marcie, PNMC, LoLA, Smart, LTSMin.
- Enriched reference tools (including the preprocessing step of [78]): LoLa+red, LTSMin+red, Marcie+red, Smart+red. These tools, since they are enriched are considered to be participants but, since the preprocessing technology they embed is issued from ITS-Tools, they are considered as a variant (i.e. only the best variant can appear in the podium of each examination). In 2020, a previous experience (called ITS-LoLA but reported as being LoLA+red in Fig. 5) already associated the 2020 version of the preprocessing phase with the 2020 version of LoLA.

---

[9] Source files are available at: https://github.com/yanntm/MCC-Drivers.

The fifteenth tool is the winner of the previous year; it is not a participating tool but a way to see how participants have evolved since the previous edition. This "virtual tool" called 2022-gold changes from examination to examination. It is tedd for StateSpace, ITS-Tools for GlobalProperties and UpperBounds, and TAPAAL for Reachability, CTL and LTL formulas. Of course, these tools are in their setting of the previous edition so that we can measure the progression compared to the winner of the previous edition, as it deals with the new benchmark.

Finally, we introduce BVT (Best Virtual Tool): the results for this tool are computed as the union of the results from all other participants. So if at least one tool answered a given query, so did the BVT.

Basics of Score Computation. The scoring follows some basic rules:

– Each examination has its own and separated scoring,
– The maximum score of the model having the largest number of instances (41) is approximatively the double of those with only one instance to limit the bias between models in the scoring,
– Since "surprise" models have never been encountered by tools, their scoring is applied a multiplier to give them a bit more importance than "known" models (even if for those, formulas are recomputed every year),
– For each miscalculated result, a penalty of twice the expected score is applied.

The scoring multiplier for the "surprise" models change regularly; they are detailed in a rule document provided at an early stage of the MCC. The rules for 2023 are presented here: https://mcc.lip6.fr/2023/pdf/rules.pdf.

The Scoring of Tools. Table 1 summarizes the results for the MCC'2023. It is divided into three parts. The first one depicts scores of competing tools. Then, we report scores from the reference tools. Finally, we show the results of 2022-gold and BVT.

Below each score line, we show the score as a percentage of the one of BVT. This is a way to outline how tools are positioned compared to BVT. We can also see the evolution compared to the winner of 2022 in each examination.

We note from Table 1 that:

– In several cases, enriched reference tools are positioned on the podium: 1st for Reachability formulas and 2nd for CTL formulas.
– The preprocessing step is quite successful by increasing the original score of reference tools. The only decrease observed is for LTSMin (-3%) for StateSpace, which is not an examination for which much improvements are expected for these techniques.
– Winners always increase their score compared to the 2022 ones. Nevertheless, 2022-gold always remain second or third place in the scoring.

Fully detailed results are available on the official web site of the MCC [60]. It details scores but also provide full results tables, as well as execution traces (when relevant), cactus plots to summarize tools' behavior, scatter plots to compare memory and time consumption, etc.

Table 1. Table depicting the score for the MCC'2023. DNC means "does not compete" (the tool does not participate in the examination) and CC means "cannot compute" (there was a bug in the driver for LTSMin which was discovered too late and no run was able to produce correct results). First ranked tool for an examination is outlined in bold-red, second in bold-green and third in **bold-blue**. Please note that ITS-Tools and <tool>+red belong to the same family, thus, only the best one among them is awarded.

	GreatSPN	smpt	Tapaal	tedd-c	ITS-Tools	LoLa+red	LTSMin+red	Marcie+red	Smart+red	LoLa	LTSMin	Marcie	pnmc	Smart	2022-gold	BVT-2023
StateSpace	14 479	DNC	DNC	16 699	12 794	DNC	8 806	11 825	10 853	DNC	9 073	11 468	10 209	9 535	16 632	17 734
	81,64%	—	—	94,16%	72,14%	—	49,66%	66,68%	61,20%	—	51,16%	64,67%	57,57%	53,77%	93,78%	100%
Global Properties	62 230	DNC	88 642	DNC	108 501	107 384	104 280	105 084	102 776	93 651	9 767	11 018	DNC	22 481	93 632	111 287
	55,92%	—	79,65%	—	97,50%	96,49%	93,70%	94,43%	92,35%	84,15%	8,78%	9,90%	—	20,20%	84,14%	100%
UpperBound	13 390	DNC	19 851	DNC	21 844	20 746	20 974	21 782	21 270	19 407	10 492	11 541	DNC	9 174	19 535	22 263
	60,15%	—	89,17%	—	98,12%	93,19%	94,21%	97,84%	95,54%	87,17%	47,13%	51,84%	—	41,21%	87,75%	100%
Reachability Formulas	21 166	42 813	44 396	DNC	44 340	44 612	43 753	44 115	43 134	39 630	22 168	19 414	DNC	16 620	44 009	45 965
	46,05%	93,14%	96,59%	—	96,46%	97,06%	95,19%	95,97%	93,84%	86,22%	48,23%	42,24%	—	36,16%	95,74%	100%
CTL Formulas	20 228	DNC	34 931	DNC	26 417	32 163	20 925	25 647	DNC	28 021	12 752	17 631	DNC	DNC	34 428	40 899
	49,46%	—	85,41%	—	64,59%	78,64%	51,16%	62,71%	—	68,51%	31,18%	43,11%	—	—	84,18%	100%
LTL Formulas	20 023	DNC	44 019	DNC	44 539	44 139	42 131	DNC	DNC	37 697	CC	DNC	DNC	DNC	44 227	45 663
	43,85%	—	96,40%	—	97,54%	96,66%	92,26%	—	—	82,55%	—	—	—	—	96,86%	100%

legend	1st	2nd	3rd

6.3 Values Computed by Tools In 2023 and Their Evolution

Evaluation of Miscalculated Values. in 2015, the MCC has introduced the notion of "Tools Confidence". It is a ratio $\frac{Correct}{Computed}$ where $Correct$ is the number of corrected values computed by the tool for the whole examination and $Computed$ the total number of values it has computed.

Since we cannot predict the results of each generated formula, we define a correct values as the one computed by a majority of at least three tools (when several variants of a given tool are submitted, they all count as one complete tool). When only one tool computes a value, it is considered as being true if its confidence is over a threshold (99,3% in 2023). Otherwise, no score is granted and the value is considered as being unknown (it is not considered as a computed one for this tool).

This notion and the associated algorithm, is considered as being quite safe. Even if, in some very rare cases, a false error is detected; it could not lead to any alteration of the scoring and ranking.

The MCC displays the confidence rate for each examination and also the global one. The "lowest" global confidence for a tool in 2023 is 99,526%[10]; it means that 574 values were miscalculated out of almost 89000 in all examinations (this tool reaches 100% in several examinations this year).

Note that confidence of participating tools has dramatically been improved since 2015.

---

[10] See https://mcc.lip6.fr/2023/results.php.

Computed Values in 2023. Table 2 shows the number of computed values by tools in 2023. The first line in each category provides absolute numbers while the second line normalizes these values compared to the "Ideal tool" which correctly computes all values for all examinations (its ratio is 100%). As in the previous table, it is separated in three parts. The first one depicts scores of competing tools. Then, we report scores from the reference tools. Finally, we show the results of 2022-gold, BVT and IdealTool.

Table 2. Table depicting the number of values computed during the MCC'2023. DNC means "does not compete" (the tool does not participate in the examination) and CC means "cannot compute" (there was a bug in the driver for LTSMin which was discovered too late and no run was able to produce correct results). Best tools (considering the number of computed values) are outlined in **bold-red**, second best tools in **bold-green** and third best tools in **bold-blue**. Please note that ITS-Tools and <tool>+red belong to the same family, thus, only the best one among them is considered.

	GreatSPN	smpt	Tapaal	tedd-c	ITS-Tools	LoLa+red	LTSMin+red	Marcie+red	Smart+red	LoLa	LTSMin	Marcie	pnmc	Smart	2022-gold	BVT-2023	Ideal tool
StateSpace	4 003	DNC	DNC	4 649	2 950	DNC	2 098	2 866	2 586	DNC	2 239	2 714	1 925	2 498	4 619	4 911	6 712
	59,64%	–	–	69,26%	43,95%	–	31,26%	42,70%	38,53%	–	33,56%	40,44%	28,68%	37,22%	66,82%	73,17%	100%
Global Properties	4 351	DNC	6 985	DNC	7 939	7 946	7 763	7 767	7 787	7 236	728	624	DNC	2 079	7 850	8 110	8 309
	51,86%	–	83,25%	–	94,62%	94,71%	92,53%	92,57%	92,81%	86,25%	8,68%	7,44%	–	24,78%	93,56%	96,66%	100%
UpperBound	14 976	DNC	23 357	DNC	25 128	24 920	24 893	25 060	24 980	22 644	11 888	10 922	DNC	9 830	24 880	25 483	26 848
	55,78%	–	87,00%	–	93,59%	92,82%	92,72%	93,34%	93,04%	84,34%	44,28%	40,68%	–	33,62%	92,67%	94,92%	100%
Reachability Formulas	22 951	50 240	50 628	DNC	50 781	52 432	50 252	50 353	49 988	44 041	24 872	18 507	DNC	17 973	50 228	52 679	53 696
	42,74%	93,56%	94,29%	–	94,57%	95,78%	93,59%	93,77%	93,09%	82,02%	46,30%	34,47%	–	33,47%	93,54%	98,11%	100%
CTL Formulas	21 759	DNC	40 462	DNC	30 509	36 414	24 000	27 184	DNC	30 422	13 587	16 337	DNC	DNC	40 043	46 421	53 696
	40,52%	–	75,35%	–	56,82%	67,82%	44,70%	50,63%	–	56,66%	25,30%	30,42%	–	–	74,57%	86,45%	100%
LTL Formulas	22 254	DNC	50 021	DNC	50 523	50 086	47 690	DNC	DNC	41 429	CC	DNC	DNC	DNC	50 303	52 012	53 696
	41,14%	–	93,16%	–	94,01%	93,28%	88,81%	–	–	77,15%	–	–	–	–	93,68%	96,83%	100%

legend	1st	2nd	3rd

We can observe that, for GlobalProperties, the tool computing the largest number of values is not the winner outlined in Table 1. There are two reasons: *(i)* the scoring puts some multipliers for "surprise models" which were never previously confronted to tools, and *(ii)*, when a tool miscalculates a value, it has a penalty of twice the expected score for this value. This may change the score when the number of computed values is very close: for GlobalProperties, ITS-Tools (winner) only computes 7 values less than Lola+red, but probably more in surprise models. However, the order of tools in the podium is correlated with the number of values computed; permutations only occurs when tools are very close in terms of performances.

We note from Table 2 that:

- As it could be noted in Table 1, the preprocessing step strongly increases the number of values computed by reference tools (up to 1 245% in the case of GlobalProperties); However, StateSpace marginally benefits from this step and LoLA, which embeds its own strategies is also less concerned.
- Winners always increase their computation capabilities compared to the 2022 ones. Nevertheless, 2022-gold always remains second or third.

- For some examinations, almost all possible values are computed (*e.g.* 95.78% for reachability formulas, and even over 96% for BVT); if this can be seen as a good improvement of tools, it also outlines that formulas complexity might be improved in the future.
- StateSpace appears to be the most difficult examination since the best tool only computed 69.26% of the values.
- BVT computes between 1.24% (UpperBound) and 14.73% more values than the best tool, showing that, if complementarity between tools exists, it remains quite low at this stage.

6.4 Evolution of the Best Tools of 2023 Since 2018 (from the Perspective of Computed Values)

Figure 5, shows, for each examination, how the three tools which compute the highest number of values in 2023 have evolved since 2018. It is a way to check how they evolved and to compare this evolution with BVT. All data is normalized to the "ideal tool" which computed all the values (representing 100% in the charts).

Let us have observations for each examination first:

- **StateSpace** (Fig. 5a): it clearly shows that over the years, the tool computing the highest number of values is very close to BVT but yet far from the "ideal tool". Since the introduction of a dedicated technique for counting states [15], TINA.tedd dominates this examination.
- **GlobalProperties** (Fig. 5b): because the properties used in this examination were changed after 2019, there is no meaningful data before 2020. In 2020, the best tool in 2023 was only capturing deadlocks but an efficient technique elaborated for deadlocks (which led to the preprocessing step associated to reference tools) was extended to other questions in 2021 and pushed forward the tool at the first place, very close to BVT. The winner follows the curve of ITS-Tools; first in the podium since 2021 (see Fig. 7) but computing 7 less values than LoLA+red.
- **UpperBound** (Fig. 5c): in 2021 the preprocessing step was also applied to this examination, thus pushing the third tool at the first place and very close to BVT. In 2018 there were 4 more participating tools and variants; this explains the large difference between the best tool and BVT.
- **Reachability formulas** (Fig. 5d): the first tool (a reference tool with the preprocessing step) holds the first position for its first participation in such a setting (as for a similar setting in 2020); LoLA, the original tool was in the podium in 2018, 2019 and 2020. The third tool's first participation was in 2021 and could capture a large number of values. It then entered in the podium at the third position in 2022 and gets very close to the other podium tools in 2023.
- **CTL formulas** (Fig. 5e): the best tool in this examination, TAPAAL, has consistently dominated the competition since 2018. It is also one of the examination where the difference between the first two tools is the largest, which may be explained by the use of specific reduction techniques. In 2018 and 2019, LoLA was very close to TAPAAL (see Fig. 7) before TAPAAL took a clear advantage.

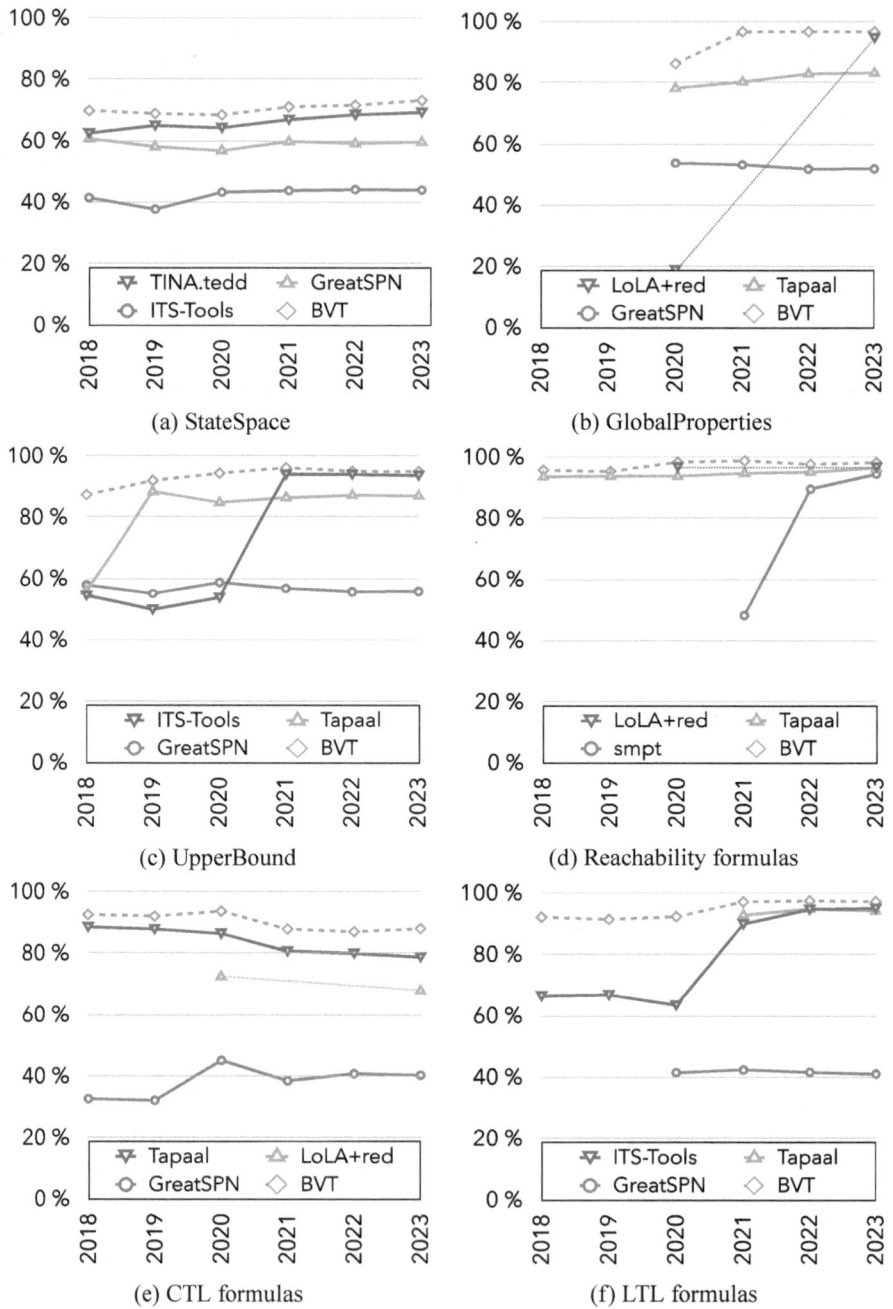

Fig. 5. Evolution of the performances of the best 2023 tools (considering the number of values computed) between 2018 and 2023. The Ideal Tool (not shown here) has always 100% of the computed values. This figure only features the three tools in the podiums for 2023 and BVT. Tools in the podium are sorted according to their rank in the MCC'2023. The line in red corresponds to the best one according to Table 2, the line in orange to the second best tool and the line in blue to the third one.

– **LTL formulas** (Fig. 5f): TAPAAL, ranked second in 2023, broke records in its first appearance in this examination, in 2021. ITS-Tools also dramatically increased its performances with the preprocessing step; it is even a bit better than TAPAAL since 2022 (a few more values computed). In 2018, for its last participation, LTS-Min was ranked second and LoLA was first in 2019. We should also underline the very good performances of EnPAC between 2020 and 2022, which was not participating in 2023 (see Fig. 7).

A recurring pattern in all the examinations is that at least one tool stays in the podium for the whole 2018–2023 period. And for an examination like StateSpace, the three podium tools have not changed. This seems to indicate that there is a bonus for the most experienced tools, which is understandable. On the other hand, we often observe that new tools rapidly reach a high position; often entering the podium. This may be explained by the fact that new tools often embed some "breakthrough technique". This occurred several times in the 2018–2023 period. For instance with the preprocessing step for GlobalProperties and UpperBound that we just described, or with a new SMT encoding for reachability properties. This also occurred twice for LTL (once with a tool that did not participate in 2023, but twice reached the third place).

We can observe a general progression of BVT in all cases except for CTL formulas. This must be tempered by the fact that there was an increase of 77.2% of the values to be computed for each examination between 2018 and 2023 (due to the introduction of new models). The decrease of BVT for CTL formulas can also be explained by the evolution in the way formulas have been generated, making them a bit more difficult. This is also true for the increase of LTL formulas where a new generation technique was introduced; it produces fairer formulas (spanning all the classes defined by Pnueli and Manna) but still needs some improvement in the way atomic propositions are selected.

We also observe that the best tool is getting closer to BVT; it means that the best tools are able to compute almost all the values computed by all the tools together, making them the really best solution.

6.5   Observations on Hardness of Examinations

With so many instances and queries, and in many cases some very high success rates, it is hard to see the specificities of each examination. That is why we proposed an innovative visualization that makes it easier to compare the complexity of each examination separately.

The plots in Fig. 6 represent the results of the examination in 2023. Each horizontal line represents a different model (with separate lines for a COL model and its unfolded PT version), so there are 158 lines in each plot. At best a line can reach the right border representing that 100% of queries were solved by the BVT in that examination for that model (all instances of it). Such lines are sorted to be at the bottom of the diagram, they are the "easiest" models. At worst a line can be empty indicating no query could be answered by any tool on any instance of that model (this is the case at the top of the state space plot for instance). Visually, if the box is full, the BVT answered all queries; the surface in white represents unsolved ones.

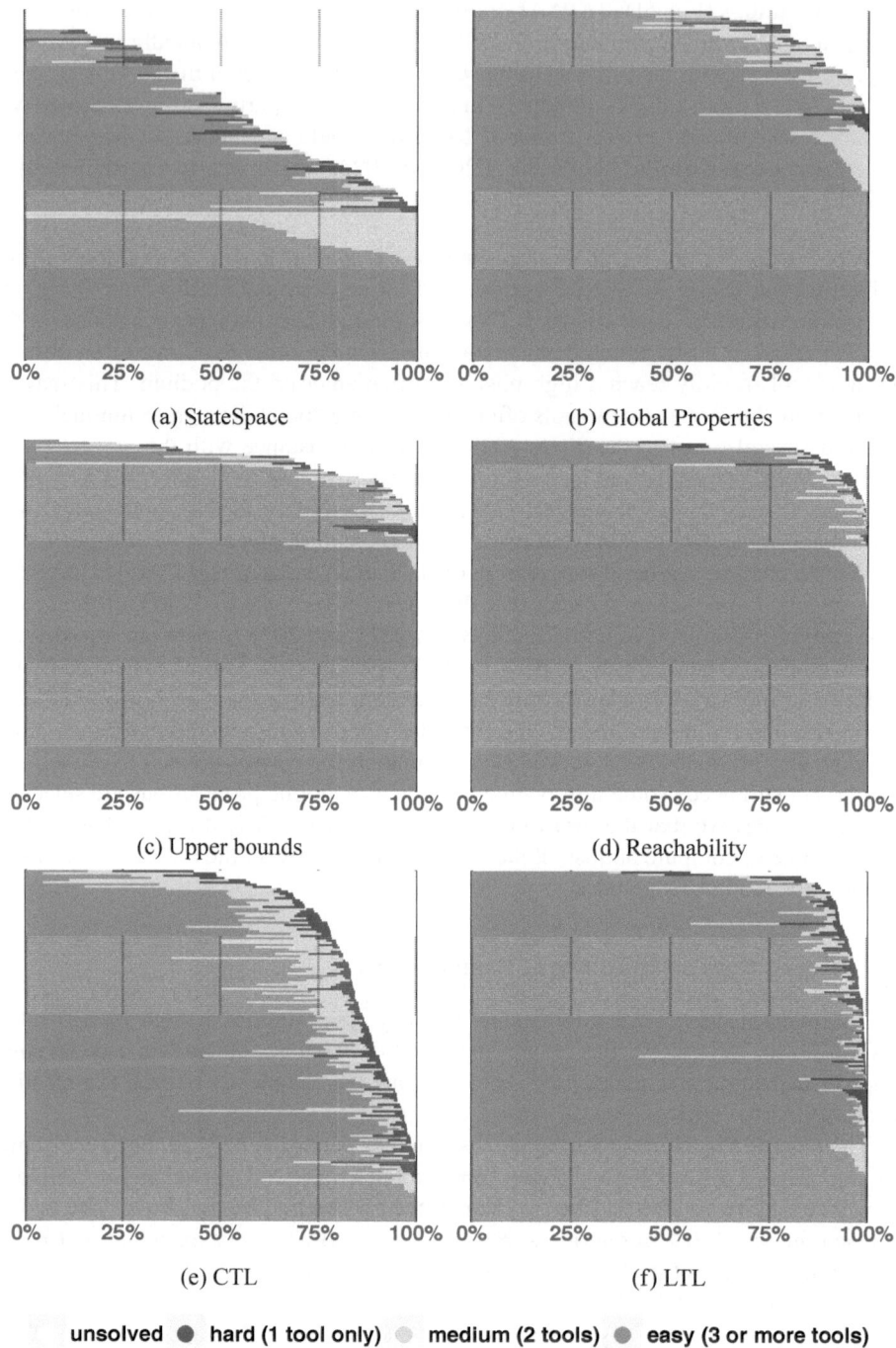

(a) StateSpace

(b) Global Properties

(c) Upper bounds

(d) Reachability

(e) CTL

(f) LTL

unsolved ● hard (1 tool only) ● medium (2 tools) ● easy (3 or more tools)

Fig. 6. Model difficulty in 2023 across different examinations.

To refine this view, the colors give an indication of tool complementarity; we count for each query that was answered how many *different* tools were able to provide that answer. Variants of a tool do not count as "different", for instance all the "+red" variants and ITS-Tools count for a single tool family. A query at least three tools answer is considered "easy", if two tools from different origins could compute the answer it is "medium", and when only one tool could compute the result it is marked as "hard". So the darker medium and hard colors correspond to existence of tool complementarity, queries that *some* tools (but not all) can solve. We designate a model as "fully solved" if the BVT could answer all queries on all model instances, and as "easy" if at least three tools could compute all queries.

- **State Space.** This category is the hardest of the MCC, with the BVT (Best Virtual Tool) only reaching 73.16% of the ideal tool score in 2023. It is also the only examination in which for some models no query could be answered. In total 70 models are fully solved (of which 43 are considered "easy"). The hardest models are identified as Echo, Election2020, FamilyReunion, HypercubeGrid, PolyORBNT, RERS. None of the queries could be answered for these models.
- **Global Properties.** This examination is currently one of the "easiest" in the MCC with the BVT reaching in percentage of "Ideal Tool": 97.55% for Reachability Deadlocks, 96.66% of Stable Marking, 100% of One Safe (NB: the only query in the contest where BVT answers for all model instances), 93.62% of Quasi Liveness and 95.47% of Liveness. Quasi-Liveness and Liveness are thus the hardest queries in this examination. In total 112 models are fully solved (of which 77 are considered "easy"). The hardest models are identified as RERS, SharedMemory (COL), DrinkVendingMachine (COL), SafeBus and EisenbergMcGuire. These models are all structurally very large.
- **Upper Bounds.** This examination is relatively easy; in 2023 the BVT reached 94.91% of the ideal score, only slightly lower than the global properties examination. In total 120 models are fully solved (of which 100 are considered "easy"). The hardest models are identified as SemanticWebServices, CANInsertWithFailure, RERS, FamilyReunion, PhilosophersDyn, FunctionPointer, Planning, VehicularWifi. Most of these models are in fact unbounded.
- **Reachability.** This examination is mostly solved by the BVT, which scores up to 98.1% of the ideal score in 2023, the highest score over all examinations. This result is quite impressive given the size of the input models (see Sect. 3) and a fortiori their state space. In total 116 models are fully solved (of which 80 are considered "easy"). The hardest models are identified as RERS, SharedMemory (COL), DoubleExponent, CANInsertWithFailure, FamilyReunion. Only the RERS models are truly hard (60% or less queries treated), 80% of the queries are answered on SharedMemory.
- **CTL.** This is the second hardest examination (after State Space) currently, the BVT scores 86.4% of the ideal score. As the colors on the plot show, there is also a significant amount of tool complementarity, with many queries solved by a single tool. The leader of the examination TAPAAL in 2023 scores 75.3% of the ideal score. In total 27 models are fully solved (of which 12 are considered "easy"). The hardest models are identified as FamilyReunion, RERS, SharedMemory (COL), Planning,

CANInsertWithFailure, Philosophers and HouseConstruction. These are mostly models for which the state space computation was not possible.

– **LTL.** Surprisingly, the LTL examination seems quite easy at the current time, with BVT reaching 96.9% of the ideal score. This is almost the level observed with Reachability. This is in part because, as discussed in Sect. 4.2, the formulas are biased towards counter-examples (70.1% of formulas can be contradicted versus only 26.7% that must be proven). In total 60 models are fully solved (of which 19 are considered "easy"), comparing these values to Reachability we see that *some* of the LTL queries are hard. The hardest models are identified as FamilyReunion, SharedMemory (COL), TokenRing, RERS, Philosophers and DatabaseWithMutex (COL). Apart from RERS, these models are mostly colored models with structurally very large unfoldings.

7 Detailed Results Analysis for 2018-2023

In this section we analyze each category of the MCC to look at the evolution of the results over the period ranging from 2018 to 2023 inclusive.

Figure 7 shows the evolution of tool performance over time as they strive to answer 100% of the queries. In these plots, the "Best Virtual Tool" BVT is computed as the union of all other tool results (hence it is always on top). To have more comparable results from year to year, all the results in this section are normalized to be a percentage of the queries that were answered. Thus 100% in the plots corresponds to answering all queries in the category. This choice deviates from the scoring used in the MCC to decide the podium, where surprise models are worth more points, errors are scored negatively, etc. But it makes the data more easily comparable on a year to year basis and is simpler to interpret.

We plot all the tools that have participated as a competitor since 2021 as well as the BVT. The tools are presented in more detail in Sect. 5 by their respective authors. The "+red" combination tools that use ITS-Tools as a preprocessing step are not represented, nor are the reference tools.

7.1 Evolution of State Space Examination

The number of participants in the examination is stable with regular contenders Tedd, GreatSPN, ITS-Tools. All these tools use symbolic decision diagram based strategies to compute the desired metrics; explicit state based methods seem to perform more poorly on this examination, and tools that use these technologies have withdrawn from participating in it over time.

ITS-Tools relies on hierarchical set decision diagrams [81] (SDD) as well as decompositions of the system using NUPN (see Sect. 3.3) and Louvain modularity [21]. Great-SPN uses the Meddly decision diagram library [13] and some advanced heuristics for variable ordering described in [10]. Tedd uses a variant of hierarchical SDD as well as a unique technology to compute these metrics on a structurally reduced net [16].

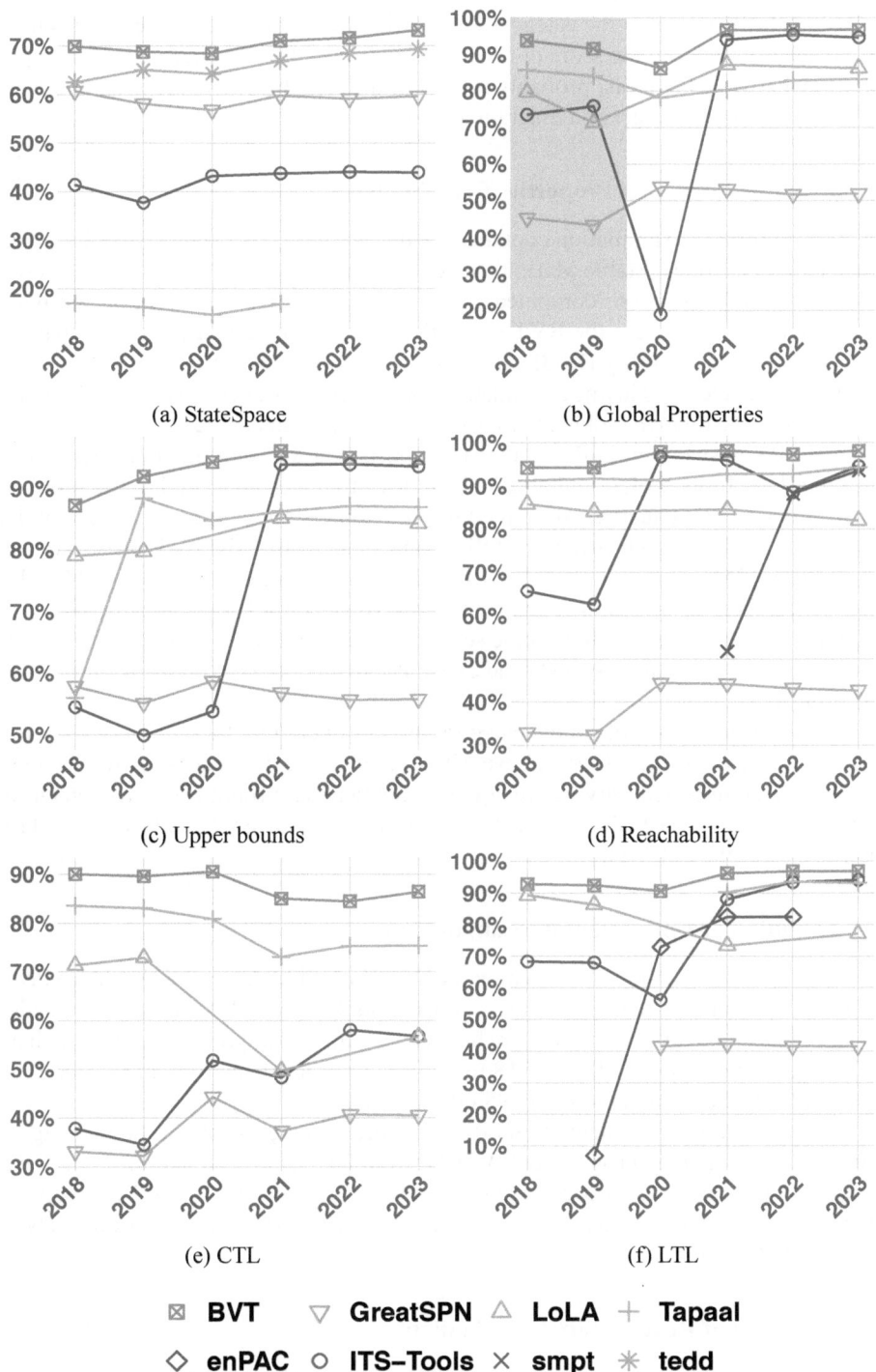

(a) StateSpace

(b) Global Properties

(c) Upper bounds

(d) Reachability

(e) CTL

(f) LTL

⊠ **BVT**   ▽ **GreatSPN**   △ **LoLA**   + **Tapaal**

◇ **enPAC**   ○ **ITS–Tools**   × **smpt**   ✳ **tedd**

Fig. 7. Evolution of tool performance across different examinations.

Tedd is the best tool in the category since 2018 where it overtook GreatSPN, and in 2023 reaches 94.66% of the score of BVT. This indicates the existence of some complementarity between the tools, probably related to the fact that static ordering heuristics for the variables in the decision diagram are critical to efficiency.

7.2 Evolution of Global Properties Examination

The global property examination contains five queries with a boolean answer expected: Reachability Deadlock, Stable Marking, One Safe, Quasi Liveness and Liveness. Before 2020, only deadlocks were computed (light gray area). This is visible on the pluri-annual plot (Fig. 7b) where the BVT score dips down in 2020, and ITS-Tools that still only supported Reachability Deadlock that year scoring below 20%.

The main contenders in this examination are ITS-Tools, LoLA, TAPAAL, GreatSPN in that order. The relative rankings of these tools is stable since 2021.

GreatSPN answers by building the state space as a decision diagram then querying it, but ITS-Tools, LoLA and TAPAAL all use strategies dedicated to the problem. LoLA uses specific reductions described in [88], some strategies dedicated to colored nets [83] as well as an advanced portfolio management [89]. TAPAAL uses advanced compression [55], structural reductions [25] and directed search heuristics [44]. ITS-Tools relies on an abstraction refinement based strategy [80] with steps dedicated to the global properties that include some semi-decision procedures, as well as both decision diagram based technology [81] and explicit state model-checking with partial order reductions [63].

Current leader ITS-Tools in the category reaches overall 97.89% of the BVT score, itself very close to the ideal score (above 95% except in Quasi Liveness). However, there is still some complementarity between the participants and room for improvement on the harder queries Quasi Liveness (ITS-Tools is at 96.3% of BVT) and Liveness (ITS-Tools is at 96.75% of BVT).

7.3 Evolution of Upper Bounds Examination

The main contenders in this examination are ITS-Tools, TAPAAL, LoLA, GreatSPN in that order. The relative rankings of these tools is stable since 2021 but the examination was introduced in MCC'2016 and there has been significant improvements in it with the BVT progressing from 87.2% in 2018 to 94.9% of the ideal score in 2023. ITS-Tools currently leads the examination with 98.60% of BVT.

Some notable improvements to tools and algorithms are visible in the plot Fig. 7c (left) such as the marked progress of TAPAAL from 2018 to 2019, continuous improvements of Lola from 2018 to 2021, and the break in performance of ITS-Tools between 2020 and 2021 due to introduction of dedicated strategies [80] instead of only relying on decision diagrams.

7.4 Evolution of Reachability Examination

The main contenders in this examination are ITS-Tools, TAPAAL, SMPT, LoLA, Great-SPN in that order. It is notable that in 2023 the top three competing tools (ITS-Tools,

TAPAAL and SMPT) are within one percent of each other at roughly 94% of the ideal score, but the BVT is much higher at 98.1% of ideal indicating that many queries could be solved by only one tool.

There has been significant improvements in this examination since 2018; the BVT score has risen from 94.2% (with leader TAPAAL in 2018 scoring 91.3% of ideal) to 98.1% in 2023 (with the leader a combination of ITS-Tools and LoLA that scored 95.8%).

Most of the current leading tools benefit from a mixture of advanced structural reduction rules to reduce the input nets, some form of linear reasoning on e.g. the state equation and guided or directed search to achieve this performance.

Notable improvements to the performance of tools are also visible on Fig. 7d, for instance the progress of GreatSPN between 2018 and 2019 due to variable ordering heuristics [10] or for ITS-Tools between 2019 and 2020 with a switch from pure decision diagrams to strategies involving SMT and structural reductions [79]. The relatively recent tool SMPT saw a drastic rise from its first submission in 2021 to become a top contender in 2023 thanks to introduction of new strategies [2]. TAPAAL also shows clear improvements from year to year with more queries and models treated, though it started from a high level in 2018 so it is less apparent on the plot. The strategies of ITS-Tools [79] to reduce the reachability problem are very effective in this examination, with all the "+red" combination tools scoring 93% or better even when the naked reference tool scores below 50%.

7.5 Evolution of CTL Examination

The main contenders in this examination are TAPAAL, LoLA, ITS-Tools and GreatSPN in that order. The BVT scores 86.4% of ideal score in 2023, which makes this the second hardest examination after state space. There is also a visible dip in the BVT score since 2021 (down to roughly 85% from roughly 90% up to 2020) that we attribute to the improvements of the formula generator Citili (see Sect. 4.1).

TAPAAL is leading the examination with 75% of queries answered, while LoLA (a reference tool in 2023) and ITS-Tools are very close to each other around 56.7%. The combination tool using ITS-Tools and LoLA performs significantly better than either in isolation at 67.8% of queries treated. GreatSPN solves 40.7% of CTL queries in 2023 but a refined analysis shows 10.6% of these answers are solved only by GreatSPN.

7.6 Evolution of LTL Examination

The main contenders in this examination are ITS-Tools, TAPAAL, EnPAC, LoLA and GreatSPN in that order. The BVT solves 97.3% of all queries a significant progression from 92% in 2018, but this high value could indicate that the formulas are not hard enough.

The leaders ITS-Tools and TAPAAL are within one percent of each other around 95%, a result that is also matched by the combination of ITS-Tools and LoLA. Then EnPAC and LoLA are around 80% of queries solved, though EnPAC did not participate in 2023 and LoLA was only submitted as a reference tool. Using its CTL* verification engine [11], GreatSPN performs similarly to CTL solving 41% of the queries.

The results of competitors in this examination have progressed a lot. TAPAAL only started supporting the LTL examination in 2021, where it immediately took the gold. EnPAC participated from 2019 where it solved 7% of queries to 2022 where it solved 82.6% of queries, an impressive progression. It obtained bronze medals in 2021 and 2022.

8   Conclusion and Novelties for the Next MCCs

We presented the Model Checking Contest and how it has been operated in its last edition (2023). We have also presented an analysis of the results for 2023, highlighted with some observations deduced from the six editions that occurred between 2018 and 2023. It shows that the MCC is an established regular event which already had a strong impact on our community: prototype tools' efficiency has been improved and a large benchmark covering models, formulas, etc., is now available.

We now sketch some novelties that could be considered to enrich the competition in the upcoming years.

8.1   Standardizing the Formula Format

This is not an issue for the competition itself but an interesting outcome for the whole community. Actually, all the participating tools have implemented libraries to parse the formulas provided in some of the examinations. Since models are provided using an ISO/IEC international standard favoring exchange and interoperability of tools, it could be of interest to provide such a standard for properties.

At this stage the format proposed in the MCC is a sort of "de facto" standard based on an XML representation of formulas. But it is not complete yet. In particular, it is not currently appropriate to be used as an exchange or storage format, but could be extended in that direction (TRUE/FALSE terminal nodes, richer atomic properties, more comparison operators than $\leq$, etc.). Thus, it needs to be refined and discussed to be possibly inserted in a future revision of the Petri net standard (ISO/IEC 15909, parts I to III).

The Model Checking Contest would help to gradually experiment with this format and improve it, so that various libraries are available when the standard is out.

8.2   Execution

At this stage, tools must report, when they provide results, the techniques they used to compute values. This could lead to an analysis of the evolution of techniques between 2015 and 2019 [62]. However, such analysis is quite complex since the vocabulary designating techniques is not normalized. Thus, manual preprocessing steps must be performed. Moreover, tools do not always differentiate the techniques for a given formula but report those of all the formulas computed in one examination.

So, there is an improvement we could complete in the next years. Then, since we get a large volume of data every year, this could rapidly enable the detection of situations were one technique (or a combination of techniques) is more efficient than other ones. Such an analysis has been proposed before [32] but the corresponding tool, competing in 2018, was not really successful.

8.3 Models

Model Forms. Each submitted model consists of one or several instances, as well as a model form. These forms, initially filled in by the people submitting the models, contain precious information about the models' origin, the methods and tools used to generate their instances, and their properties. The model board members carefully check the content of these forms, enriching them as much as possible with the results provided by the tools participating in the MCC, as well as other tools (e.g. CÆSAR.BDD[11] and ConcNUPN [29]). This is a difficult and time-consuming task, particularly for models including instances that cannot be handled by state-of-the-art tools, or presenting unexpected specifics (e.g. divergent behaviors between colored and PT instances). It happens quite regularly that models that are several years old need to be enhanced.

Currently, model forms (in tex format) are processed to produce XML files[12] giving, for each model, a total of 24 global properties (each of which can be evaluated as true, false or unknown). Currently, when there are divergences among instances of a model (for example, where some instances are safe, while others are not), the property has to be marked as "unknown" and we have to write manually a statement that is only readable by humans. We would like to improve this format, to be able to describe these properties for each instance, possibly with new useful properties for the community, and possibly beyond ternary evaluations (for example, it is helpful to know the names of (all) dead transitions, or at least their number, rather than merely knowing their existence).

Correction of Existing Models. As far as possible, we avoid altering instances from previous editions of the MCC, the only exception being when they present serious, unnoticed problems. For example, instances containing places or transitions with label names different from their XML id, a nice feature of the PNML format, but leading to incorrect formula evaluation by certain tools not taking this feature into account. Or when some instances are found to be isomorphic (there is a permutation of their places, their transitions and, in the case of NUPNs, their units, such that these two instances coincide): in this case, we eliminate the duplicate.

Model Enhancements. In order to follow the noticeable progression of model-checkers through successive editions and to keep existing models "challenging" for these tools, we can add extra instances to existing models.

Feeding Some Examinations with Structural Information. One idea should be to feed the reachability, CTL and LTL examinations with structural information (extracted when computed from GlobalProperties questions). This information can help tools to improve their algorithms.

New Petri Net Formalisms (e.g. Time). So far, we support Petri nets and colored (*i.e.* Symmetric) Petri nets; both are described in the ISO/IEC standard. One suggestion is

---

[11] https://cadp.inria.fr/man/caesar.bdd.html.

[12] https://mcc.lip6.fr/verdict-properties.php.

to extend the MCC to time(d) nets. This is a difficult task since there are only a few tools dealing with time in Petri nets. Moreover, these tools support different timing schemes. The main ones are: Time Petri Nets [68] that associate a firing interval with each transition; Timed Petri Nets [72] which feature a global clock and tokens carry an availability time; and Timed-Arc Petri Nets [47] where tokens carry an age and arcs between places and transitions are labelled with time intervals restricting the age of tokens available for transition firing.

The idea is to define the core semantics supported by existing tools and to encode it using the new extensions mechanisms proposed in the Part III of the ISO/IEC standard. Integrating such Petri Nets extensions must involve both the tool developers and the MCC organizers.

References

1. Amat, N.: Octant: The Reachability Formula Projector. A tool to project Petri net reachability properties on reduced nets using polyhedral reduction (2023). https://github.com/nicolasAmat/Octant
2. Amat, N., Berthomieu, B., Dal Zilio, S.: A polyhedral abstraction for petri nets and its application to SMT-based model checking. Fundamenta Informaticae **187**(2-4) (2022). https://doi.org/10.3233/FI-222134
3. Amat, N., Chauvet, L.: Kong: a tool to squash concurrent places. In: Bernardinello, L., Petrucci, L. (eds.) PETRI NETS 2022. LNCS, vol. 13288, pp. 115–126. Springer, Cham (2022). https://doi.org/10.1007/978-3-031-06653-5_6
4. Amat, N., Dal Zilio, S.: SMPT: a testbed for reachability methods in generalized Petri nets. In: Chechik, M., Katoen, J.P., Leucker, M. (eds.) FM 2023. LNCS, vol. 14000, pp. 445–453. Springer, Cham (2023). https://doi.org/10.1007/978-3-031-27481-7_25
5. Amat, N., Zilio, S.D., Hujsa, T.: Property directed reachability for generalized petri nets. In: TACAS 2022. LNCS, vol. 13243, pp. 505–523. Springer, Cham (2022). https://doi.org/10.1007/978-3-030-99524-9_28
6. Amat, N., Dal Zilio, S., Le Botlan, D.: Leveraging polyhedral reductions for solving Petri net reachability problems. Int. J. Softw. Tools Technol. Transfer (2022). https://doi.org/10.1007/s10009-022-00694-8
7. Amat, N., Dal Zilio, S., Le Botlan, D.: Automated polyhedral abstraction proving. In: Gomes, L., Lorenz, R. (eds.) PETRI NETS 2023. LNCS, vol. 13929, pp. 324–345. Springer, Cham (2023). https://doi.org/10.1007/978-3-031-33620-1_18
8. Amparore, E.G.: Stochastic modelling and evaluation using GreatSPN. ACM SIGMETRICS Perform. Eval. Rev. **49**(4), 87–91 (2022)
9. Amparore, E.G., Ciardo, G., Miner, A.S.: The footprint form of a matrix: definition, properties, and an application. Linear Algebra Appl. **651**, 209–229 (2022)
10. Amparore, E.G., Donatelli, S., Ciardo, G.: Variable order metrics for decision diagrams in system verification. Int. J. Softw. Tools Technol. Transf. **22**(5), 541–562 (2020)
11. Amparore, E.G., Donatelli, S., Gallà, F.: starMC: an automata based CTL* model checker. PeerJ Comput. Sci. **8**, e823 (2022)
12. Andersen, M., Gatten Larsen, H., Srba, J., Grund Sørensen, M., Haahr Taankvist, J.: Verification of liveness properties on closed timed-arc petri nets. In: Kučera, A., Henzinger, T.A., Nešetřil, J., Vojnar, T., Antoš, D. (eds.) MEMICS 2012. LNCS, vol. 7721, pp. 69–81. Springer, Heidelberg (2013). https://doi.org/10.1007/978-3-642-36046-6_8
13. Babar, J., Miner, A.S.: Meddly: multi-terminal and edge-valued decision diagram library. In: QEST, pp. 195–196. IEEE Computer Society (2010)

14. Bérard, B., et al.: Systems and Software Verification, Model-Checking Techniques and Tools. Springer, Heidelberg (2001). https://doi.org/10.1007/978-3-662-04558-9
15. Berthomieu, B., Le Botlan, D., Dal Zilio, S.: Petri net reductions for counting markings. In: Gallardo, M.M., Merino, P. (eds.) SPIN 2018. LNCS, vol. 10869, pp. 65–84. Springer, Cham (2018). https://doi.org/10.1007/978-3-319-94111-0_4
16. Berthomieu, B., Le Botlan, D., Dal Zilio, S.: Counting petri net markings from reduction equations. Int. J. Softw. Tools Technol. Transfer **22** (2019). https://doi.org/10.1007/s10009-019-00519-1
17. Berthomieu, B., Ribet, P.O., Vernadat, F.: The tool TINA-construction of abstract state spaces for Petri nets and time Petri nets. Int. J. Prod. Res. **42**(14), 2741–2756 (2004). https://doi.org/10.1080/00207540412331312688
18. Berthomieu, B., Vernadat, F.: Time petri nets analysis with TINA. In: QEST, pp. 123–124. IEEE Computer Society (2006)
19. Bilgram, A., Jensen, P.G., Pedersen, T., Srba, J., Taankvist, P.H.: Improvements in unfolding of colored petri nets. In: Bell, P.C., Totzke, P., Potapov, I. (eds.) RP 2021. LNCS, vol. 13035, pp. 69–84. Springer, Cham (2021). https://doi.org/10.1007/978-3-030-89716-1_5
20. Bilgram, A., Jensen, P., Pedersen, T., Srba, J., Taankvist, P.: Methods for efficient unfolding of colored petri nets. Fundamenta Informaticae 1–24 (2023, to appear)
21. Blondel, V.D., Guillaume, J., Lambiotte, R., Lefebvre, E.: Fast unfolding of community hierarchies in large networks. CoRR abs/0803.0476 (2008)
22. Bobot, F., Bromberger, M., Hoenicke, J.: The International SMT Competition Web Page (2023). https://smt-comp.github.io/
23. Bønneland, F., Dyhr, J., Jensen, P.G., Johannsen, M., Srba, J.: Simplification of CTL formulae for efficient model checking of petri nets. In: Khomenko, V., Roux, O.H. (eds.) PETRI NETS 2018. LNCS, vol. 10877, pp. 143–163. Springer, Cham (2018). https://doi.org/10.1007/978-3-319-91268-4_8
24. Bønneland, F.M., Jensen, P.G., Larsen, K.G., Muñiz, M., Srba, J.: Stubborn set reduction for timed reachability and safety games. In: Dima, C., Shirmohammadi, M. (eds.) FORMATS 2021. LNCS, vol. 12860, pp. 32–49. Springer, Cham (2021). https://doi.org/10.1007/978-3-030-85037-1_3
25. Bønneland, F., Dyhr, J., Jensen, P., Johannsen, M., Srba, J.: Stubborn versus structural reductions for petri nets. J. Log. Algebraic Methods Program. **102**(1), 46–63 (2019)
26. Bønneland, F., Jensen, P., Larsen, K., Muniz, M., Srba, J.: Partial order reduction for reachability games. In: Proceedings of the 30th International Conference on Concurrency Theory (CONCUR 2019). LIPICS, vol. 140, pp. 23:1–23:15. Dagstuhl Publishing (2019)
27. Bouvier, P.: The VLSAT-3 Benchmark Suite. Technical report RT-0516, INRIA, Grenoble, France (2021). https://hal.inria.fr/hal-3468625. https://arxiv.org/abs/2112.03675
28. Bouvier, P., Garavel, H.: The VLSAT-1 Benchmark Suite. Technical report RT-0510, INRIA, Grenoble, France (2020). https://hal.inria.fr/hal-03007233. https://arxiv.org/abs/2011.11049
29. Bouvier, P., Garavel, H.: Efficient algorithms for three reachability problems in safe petri nets. In: Buchs, D., Carmona, J. (eds.) PETRI NETS 2021. LNCS, vol. 12734, pp. 339–359. Springer, Cham (2021). https://doi.org/10.1007/978-3-030-76983-3_17
30. Bouvier, P., Garavel, H.: The VLSAT-2 Benchmark Suite. Technical report RT-0514, INRIA, Grenoble, France (2021). https://hal.inria.fr/hal-03337115. https://arxiv.org/abs/2110.06336
31. Bouvier, P., Garavel, H., Ponce-de-León, H.: Automatic decomposition of petri nets into automata networks – a synthetic account. In: Janicki, R., Sidorova, N., Chatain, T. (eds.) PETRI NETS 2020. LNCS, vol. 12152, pp. 3–23. Springer, Cham (2020). https://doi.org/10.1007/978-3-030-51831-8_1

32. Buchs, D., Klikovits, S., Linard, A., Mencattini, R., Racordon, D.: A model checker collection for the model checking contest using docker and machine learning. In: Khomenko, V., Roux, O.H. (eds.) PETRI NETS 2018. LNCS, vol. 10877, pp. 385–395. Springer, Cham (2018). https://doi.org/10.1007/978-3-319-91268-4_21
33. Byg, J., Jørgensen, K.Y., Srba, J.: An efficient translation of timed-arc petri nets to networks of timed automata. In: Breitman, K., Cavalcanti, A. (eds.) ICFEM 2009. LNCS, vol. 5885, pp. 698–716. Springer, Heidelberg (2009). https://doi.org/10.1007/978-3-642-10373-5_36
34. Byg, J., Jørgensen, K.Y., Srba, J.: TAPAAL: editor, simulator and verifier of timed-arc petri nets. In: Liu, Z., Ravn, A.P. (eds.) ATVA 2009. LNCS, vol. 5799, pp. 84–89. Springer, Heidelberg (2009). https://doi.org/10.1007/978-3-642-04761-9_7
35. Chiola, G., Dutheillet, C., Franceschinis, G., Haddad, S.: A symbolic reachability graph for coloured Petri nets. Theoret. Comput. Sci. **176**(1–2), 39–65 (1997)
36. Christensen, S., Kristensen, L.M., Mailund, T.: A sweep-line method for state space exploration. In: Margaria, T., Yi, W. (eds.) TACAS 2001. LNCS, vol. 2031, pp. 450–464. Springer, Heidelberg (2001). https://doi.org/10.1007/3-540-45319-9_31
37. Dal Zilio, S.: MCC: a tool for unfolding colored petri nets in PNML format. In: Janicki, R., Sidorova, N., Chatain, T. (eds.) PETRI NETS 2020. LNCS, vol. 12152, pp. 426–435. Springer, Cham (2020). https://doi.org/10.1007/978-3-030-51831-8_23
38. Dalsgaard, A., et al.: A distributed fixed-point algorithm for extended dependency graphs. Fund. Inform. **161**(4), 351–381 (2018)
39. Dalsgaard, A.E., et al.: Extended dependency graphs and efficient distributed fixed-point computation. In: van der Aalst, W., Best, E. (eds.) PETRI NETS 2017. LNCS, vol. 10258, pp. 139–158. Springer, Cham (2017). https://doi.org/10.1007/978-3-319-57861-3_10
40. Dalsgaard, A.E., Enevoldsen, S., Larsen, K.G., Srba, J.: Distributed computation of fixed points on dependency graphs. In: Fränzle, M., Kapur, D., Zhan, N. (eds.) SETTA 2016. LNCS, vol. 9984, pp. 197–212. Springer, Cham (2016). https://doi.org/10.1007/978-3-319-47677-3_13
41. David, A., Jacobsen, L., Jacobsen, M., Jørgensen, K.Y., Møller, M.H., Srba, J.: TAPAAL 2.0: integrated development environment for timed-arc petri nets. In: Flanagan, C., König, B. (eds.) TACAS 2012. LNCS, vol. 7214, pp. 492–497. Springer, Heidelberg (2012). https://doi.org/10.1007/978-3-642-28756-5_36
42. David, A., Jacobsen, L., Jacobsen, M., Srba, J.: A forward reachability algorithm for bounded timed-arc Petri nets. In: Proceedings of the 7th International Conference on Systems Software Verification (SSV 2012). EPTCS, vol. 102, pp. 125–140. Open Publishing Association (2012)
43. Duret-Lutz, A., et al.: From spot 2.0 to spot 2.10: what's new? In: Shoham, S., Vizel, Y. (eds.) CAV 2022. LNCS, vol. 13372, pp. 174–187. Springer, Cham (2022). https://doi.org/10.1007/978-3-031-13188-2_9
44. Henriksen, E.G., et al.: Potency-based heuristic search with randomness for explicit model checking. In: Caltais, G., Schilling, C. (eds.) SPIN 2023. LNCS, vol. 13872, pp. 180–187. Springer, Cham (2023). https://doi.org/10.1007/978-3-031-32157-3_10
45. Garavel, H.: Nested-unit petri nets. J. Log. Algebraic Methods Program. **104**, 60–85 (2019)
46. Geldenhuys, J., Valmari, A.: More efficient on-the-fly LTL verification with Tarjan's algorithm. Theor. Comput. Sci. **345**(1), 60–82 (2005)
47. Hanisch, H.-M.: Analysis of place/transition nets with timed arcs and its application to batch process control. In: Ajmone Marsan, M. (ed.) ICATPN 1993. LNCS, vol. 691, pp. 282–299. Springer, Heidelberg (1993). https://doi.org/10.1007/3-540-56863-8_52
48. He, C., Ding, Z.: More efficient on-the-fly verification methods of colored petri nets. Comput. Inform. **40**(1), 195–215 (2021)
49. Hecher, M., Fichte, J.: The Model Counting Competition Web Page (2023). https://mccompetition.org

50. Heule, M., Jarvisalo, M., Suda, M., Iser, M., Balyo, T.: The International SAT Competition Web Page (2023). http://www.satcompetition.org/
51. Hillah, L., Kordon, F., Lakos, C., Petrucci, L.: Extending PNML scope: the prioritised petri nets experience. In: Petri Net and Software Engineering (PNSE 2011), Newcastle, UK, vol. 723, pp. 61–75. CEUR (2011)
52. Jacobsen, L., Jacobsen, M., Møller, M.H., Srba, J.: A framework for relating timed transition systems and preserving TCTL model checking. In: Aldini, A., Bernardo, M., Bononi, L., Cortellessa, V. (eds.) EPEW 2010. LNCS, vol. 6342, pp. 83–98. Springer, Heidelberg (2010). https://doi.org/10.1007/978-3-642-15784-4_6
53. Jensen, J., Nielsen, T., Oestergaard, L., Srba, J.: TAPAAL and reachability analysis of P/T nets. LNCS Trans. Petri Nets Other Models Concurr. (ToPNoC) **9930**, 307–318 (2016)
54. Jensen, P.G., Larsen, K.G., Srba, J.: Real-time strategy synthesis for timed-arc petri net games via discretization. In: Bošnački, D., Wijs, A. (eds.) SPIN 2016. LNCS, vol. 9641, pp. 129–146. Springer, Cham (2016). https://doi.org/10.1007/978-3-319-32582-8_9
55. Jensen, P., Larsen, K., Srba, J.: PTrie: data structure for compressing and storing sets via prefix sharing. In: Hung, D., Kapur, D. (eds.) ICTAC 2017. LNCS, vol. 10580, pp. 248–265. Springer, Cham (2017). https://doi.org/10.1007/978-3-319-67729-3_15
56. Jensen, P.G., Larsen, K.G., Srba, J., Sørensen, M.G., Taankvist, J.H.: Memory efficient data structures for explicit verification of timed systems. In: Badger, J.M., Rozier, K.Y. (eds.) NFM 2014. LNCS, vol. 8430, pp. 307–312. Springer, Cham (2014). https://doi.org/10.1007/978-3-319-06200-6_26
57. Jensen, P., Larsen, K., Srba, J., Ulrik, N.: Elimination of detached regions in dependency graph verification. In: Caltais, G., Schilling, C. (eds.) SPIN 2023. LNCS, vol. 13872, pp. 163–179. Springer, Cham (2023). https://doi.org/10.1007/978-3-031-32157-3_9
58. Jensen, P.G., Srba, J., Ulrik, N.J., Virenfeldt, S.M.: Automata-driven partial order reduction and guided search for LTL model checking. In: Finkbeiner, B., Wies, T. (eds.) VMCAI 2022. LNCS, vol. 13182, pp. 151–173. Springer, Cham (2022). https://doi.org/10.1007/978-3-030-94583-1_8
59. Kant, G., Laarman, A., Meijer, J., van de Pol, J., Blom, S., van Dijk, T.: LTSmin: high-performance language-independent model checking. In: Baier, C., Tinelli, C. (eds.) TACAS 2015. LNCS, vol. 9035, pp. 692–707. Springer, Heidelberg (2015). https://doi.org/10.1007/978-3-662-46681-0_61
60. Kordon, F., et al.: Complete Results for the 2023 Edition of the Model Checking Contest (2023). https://mcc.lip6.fr/2023/results.php
61. Kordon, F., Linard, A., Paviot-Adet, E.: Optimized colored nets unfolding. In: Najm, E., Pradat-Peyre, J.-F., Donzeau-Gouge, V.V. (eds.) FORTE 2006. LNCS, vol. 4229, pp. 339–355. Springer, Heidelberg (2006). https://doi.org/10.1007/11888116_25
62. Kordon, F., Hillah, L., Hulin-Hubard, F., Jezequel, L., Paviot-Adet, E.: Study of the efficiency of model checking techniques using results of the MCC from 2015 to 2019. Int. J. Softw. Tools Technol. Transf. **23**(6), 931–952 (2021)
63. Laarman, A.: Stubborn transaction reduction. In: Dutle, A., Muñoz, C., Narkawicz, A. (eds.) NFM 2018. LNCS, vol. 10811, pp. 280–298. Springer, Cham (2018). https://doi.org/10.1007/978-3-319-77935-5_20
64. Lehmann, A., Lohmann, N., Wolf, K.: Stubborn sets for simple linear time properties. In: Haddad, S., Pomello, L. (eds.) PETRI NETS 2012. LNCS, vol. 7347, pp. 228–247. Springer, Heidelberg (2012). https://doi.org/10.1007/978-3-642-31131-4_13
65. Liebke, T., Wolf, K.: Taking some burden off an explicit CTL model checker. In: Donatelli, S., Haar, S. (eds.) PETRI NETS 2019. LNCS, vol. 11522, pp. 321–341. Springer, Cham (2019). https://doi.org/10.1007/978-3-030-21571-2_18

66. Manna, Z., Pnueli, A.: A hierarchy of temporal properties. In: Proceedings of the 9th ACM Symposium on Principles of Distributed Computing (PODC 1990), pp. 377–410. ACM Press (1990)
67. Mateo, J., Srba, J., Sørensen, M.: Soundness of timed-arc workflow nets in discrete and continuous-time semantics. Fund. Inform. **140**(1), 89–121 (2015)
68. Merlin, P.M., Farber, D.J.: Recoverability of communication protocols-implications of a theoretical study. IEEE Trans. Commun. **24**(9), 1036–1043 (1976)
69. de Moura, L., Bjørner, N.: Z3: an efficient SMT solver. In: Ramakrishnan, C.R., Rehof, J. (eds.) TACAS 2008. LNCS, vol. 4963, pp. 337–340. Springer, Heidelberg (2008). https://doi.org/10.1007/978-3-540-78800-3_24
70. Paviot-Adet, E., Poitrenaud, D., Renault, E., Thierry-Mieg, Y.: LTL under reductions with weaker conditions than stutter invariance. In: Mousavi, M.R., Philippou, A. (eds.) FORTE 2022. LNCS, vol. 13273, pp. 170–187. Springer, Cham (2022). https://doi.org/10.1007/978-3-031-08679-3_11
71. Peterson, G.L.: Myths about the mutual exclusion problem. Inf. Process. Lett. **12**(3), 115–116 (1981)
72. Ramchandani, C.: Analysis of asynchronous concurrent systems by timed Petri nets. Ph.D. thesis, MIT, USA (1973)
73. Schmidt, K.: Model-checking with coverability graphs. Formal Methods Syst. Des. **15**(3), 239–254 (1999)
74. Schmidt, K.: Stubborn sets for standard properties. In: Donatelli, S., Kleijn, J. (eds.) ICATPN 1999. LNCS, vol. 1639, pp. 46–65. Springer, Heidelberg (1999). https://doi.org/10.1007/3-540-48745-X_4
75. Schmidt, K.: How to calculate symmetries of petri nets. Acta Informatica **36**(7), 545–590 (2000)
76. Schmidt, K.: Automated generation of a progress measure for the sweep-line method. In: Jensen, K., Podelski, A. (eds.) TACAS 2004. LNCS, vol. 2988, pp. 192–204. Springer, Heidelberg (2004). https://doi.org/10.1007/978-3-540-24730-2_17
77. Thierry-Mieg, Y.: Symbolic model-checking using ITS-tools. In: Baier, C., Tinelli, C. (eds.) TACAS 2015. LNCS, vol. 9035, pp. 231–237. Springer, Heidelberg (2015). https://doi.org/10.1007/978-3-662-46681-0_20
78. Thierry-Mieg, Y.: Structural reductions revisited. In: Janicki, R., Sidorova, N., Chatain, T. (eds.) PETRI NETS 2020. LNCS, vol. 12152, pp. 303–323. Springer, Cham (2020). https://doi.org/10.1007/978-3-030-51831-8_15
79. Thierry-Mieg, Y.: Symbolic and structural model-checking. Fundam. Informaticae **183**(3–4), 319–342 (2021)
80. Thierry-Mieg, Y.: Efficient strategies to compute invariants, bounds and stable places of petri nets. In: PNSE @ Petri Nets. CEUR Workshop Proceedings, vol. TBA, pp. 16–33. CEUR-WS.org (2023)
81. Thierry-Mieg, Y., Poitrenaud, D., Hamez, A., Kordon, F.: Hierarchical set decision diagrams and regular models. In: Kowalewski, S., Philippou, A. (eds.) TACAS 2009. LNCS, vol. 5505, pp. 1–15. Springer, Heidelberg (2009). https://doi.org/10.1007/978-3-642-00768-2_1
82. Valmari, A.: Stubborn sets for reduced state space generation. In: Rozenberg, G. (ed.) ICATPN 1989. LNCS, vol. 483, pp. 491–515. Springer, Heidelberg (1991). https://doi.org/10.1007/3-540-53863-1_36
83. Wallner, S., Wolf, K.: Skeleton abstraction for universal temporal properties. Fundam. Informaticae **187**(2–4), 245–272 (2022)
84. Wimmel, H., Wolf, K.: Applying CEGAR to the petri net state equation. Log. Methods Comput. Sci. **8**(3) (2012)
85. Wolf, K.: The petri net twist in explicit model checking. Softw. Syst. Model. **14**(2), 711–717 (2015)

86. Wolf, K.: Running lola 2.0 in a model checking competition. Trans. Petri Nets Other Model. Concurr. **11**, 274–285 (2016)
87. Wolf, K.: Petri net model checking with LoLA 2. In: Khomenko, V., Roux, O.H. (eds.) PETRI NETS 2018. LNCS, vol. 10877, pp. 351–362. Springer, Cham (2018). https://doi.org/10.1007/978-3-319-91268-4_18
88. Wolf, K.: How petri net theory serves petri net model checking: a survey. Trans. Petri Nets Other Model. Concurr. **14**, 36–63 (2019)
89. Wolf, K.: Portfolio management in explicit model checking. Trans. Petri Nets Other Model. Concurr. **16**, 91–111 (2022)

Tools at the Frontiers of Quantitative Verification
QComp 2023 Competition Report

Roman Andriushchenko[1], Alexander Bork[2], Carlos E. Budde[3],
Milan Češka[1], Kush Grover[4], Ernst Moritz Hahn[5],
Arnd Hartmanns[5]([✉]), Bryant Israelsen[6], Nils Jansen[7],
Joshua Jeppson[6], Sebastian Junges[7], Maximilian A. Köhl[8],
Bettina Könighofer[9], Jan Křetínský[4,10], Tobias Meggendorfer[4,11,12],
David Parker[13], Stefan Pranger[9], Tim Quatmann[2], Enno Ruijters[5],
Landon Taylor[6], Matthias Volk[14], Maximilian Weininger[4,11],
and Zhen Zhang[6]

[1] Brno University of Technology, Brno, Czech Republic
[2] RWTH Aachen University, Aachen, Germany
[3] University of Trento, Trento, Italy
[4] Technical University of Munich, Munich, Germany
[5] University of Twente, Enschede, The Netherlands
`a.hartmanns@utwente.nl`
[6] Utah State University, Logan, UT, USA
[7] Radboud University, Nijmegen, The Netherlands
[8] Saarland University, Saarland Informatics Campus, Saarbrücken, Germany
[9] Graz University of Technology, Graz, Austria
[10] Masaryk University, Brno, Czech Republic
[11] Institute of Science and Technology Austria, Klosterneuburg, Austria
[12] Lancaster University Leipzig, Leipzig, Germany
[13] University of Oxford, Oxford, UK
[14] Eindhoven University of Technology, Einhoven, The Netherlands

Abstract. The analysis of formal models that include quantitative aspects such as timing or probabilistic choices is performed by quantitative verification tools. Broad and mature tool support is available for computing basic properties such as expected rewards on basic models such as Markov chains. Previous editions of QComp, the comparison of tools for the analysis of quantitative formal models, focused on this setting. Many application scenarios, however, require more advanced property types such as LTL and parameter synthesis queries as well as

The authors are ordered alphabetically. This work was supported by DFG RTG 2236/2 (UnRAVeL) and DFG project TRR 248 (CPEC, ID 389792660), by the EU under MSCA grant agreements 101008233 (MISSION), 101034413 (IST-BRIDGE), and 101067199 (ProSVED), by ERC Starting Grant 101077178 (DEUCE), ERC Consolidator Grant 864075 (CAESAR), and ERC Advanced Grant 834115 (FUN2MODEL), by GAČR grant GA23-06963S (VESCAA), by National Science Foundation grant 1856733, by NextGenerationEU project D53D23008400006 (SMARTITUDE), and by NWO VENI grant 639.021.754.

advanced models like stochastic games and partially observable MDPs. For these, tool support is in its infancy today. This paper presents the outcomes of QComp 2023: a survey of the state of the art in quantitative verification tool support for advanced property types and models. With tools ranging from first research prototypes to well-supported integrations into established toolsets, this report highlights today's active areas and tomorrow's challenges in tool-focused research for quantitative verification.

1 Introduction

The inclusion of quantitative aspects such as probabilistic choices, timing, and random delays in system modelling is crucial to ensure the correctness, performance, and dependability of the ever-increasing amount of complex safety- and economically-critical systems that support our societies. Well-known examples include the use of randomised algorithms in Internet protocols to achieve both simplicity and scalability [156] or the fault tree modelling approach for safety assessment in the nuclear industry [94].

Formally, these aspects can be captured in established mathematical **formalisms** like discrete- and continuous-time Markov chains (DTMCs and CTMCs) for probabilistic choices and stochastic timing, or more recent notions such as timed automata (TA) [4] for real-time behaviour. Combining DTMCs with nondeterministic (i.e. unquantified and controllable or adversarial) choices results in the nowadays-popular formalism of Markov decision processes (MDPs) [29,194]. These form the mathematical foundation of *quantitative modelling*; for practical purposes, models are specified in a higher-level **modelling language**—such as MODEST [30,105] or the PRISM language [163]—that is equipped with a semantics in terms of one of the formalisms. When combined with a query for a numerical *property* of a model, e.g. for the probability of reaching a set of undesirable states or for the expected reward until a terminal state is reached, we have a basic *quantitative verification* problem.

Basic Quantitative Verification Comparisons

The *basic problems*—i.e. computing a (i) reachability probability, (ii) expected accumulated reward, or (iii) steady-state probability on a DTMC, CTMC, or MDP model[1]—can be solved by various software tools developed over the past two decades. Most tools use one of two approaches: either probabilistic model checking (PMC) [20,116], which applies a numeric algorithm onto a complete in-memory representation of a model's state space, or statistical model checking (SMC) [3,169,221], which randomly samples (or: *simulates*) and statistically analyses a set of model behaviours; or a hybrid approach combining aspects of PMC and SMC such as partial exploration [148], probabilistic planning [142,176], (deep) reinforcement learning [36,98], or Monte Carlo tree search [11].

---

[1] Probabilistic timed automata (PTA) [166] can be turned into equivalent MDP [165, 167] (or be solved as stochastic games [162]), so treat them like MDP here.

The QComp Competition. The 2019 Comparison of Tools for the Analysis of Quantitative Formal Models (QComp 2019) [103] compared the performance, versatility, and usability of nine such tools on a benchmark set of 100 basic quantitative verification problems[2]. It was the first tool competition in quantitative verification, part of the TOOLympics at TACAS 2019 [25]. The next edition of QComp in 2020 [46] used the same benchmark set, but focused more specifically on the different types of correctness guarantees provided by the different tools, highlighting the interplay between performance and precision in quantitative verification. The results of QComp 2020 were presented at the ISoLA 2020/2021 conference. Although the main outcomes of these two editions of QComp were performance results, they were meant as *friendly competitions*: We did not establish a ranking of tools or point out a "winner"; rather, we highlighted the capabilities, strengths, and specific niches of all participating tools. In particular, the results clearly showed that some tools were generalists solving many types of problems, while others were specialised to specific tasks where they performed much better than any other participant. The entire performance evaluation and report-writing process was performed in close collaboration with the participants, most of which were the main developers of the respective tools.

Benchmark Sets and Formats. Aside from providing information about the capabilities and performance of the participating tools, these two editions of QComp also benefited the collaboration and alignment inside the quantitative verification research community: In a parallel effort to QComp 2019, we established the Quantitative Verification Benchmark Set (QVBS) [119], from which the competition selected its 100 benchmark instances. Although the QVBS' models were collected from various sources and came in various modelling languages, the QVBS as a matter of principle includes a translation of each model and its properties into the JANI interchange format [44]; as a result, any tool that supported JANI could participate in QComp 2019 and 2020. JANI thus benefited QComp and the participating tool authors by simplifying frontend development, while QComp furthered the establishment of JANI as a community standard.

QComp 2023: Looking into the Future

While improving solution methods for basic problems remains an active research topic (cf. e.g. [28,102,111,116,117,128]), most of today's work in quantitative verification focuses on what we refer to as *advanced problems*: Computing more complex properties on the basic models, computing basic properties on more complex models, or combinations thereof. Most papers include an experimental evaluation, which, however, often uses an ad-hoc research prototype implementation, most of which are *not* further developed into a stable and maintained

---

[2] In the tool competition context, our verification problems are called *benchmark instances*. Since most benchmark models are parametrised but basic problems ask for a single result value, a benchmark instance is a triple of a model, a concrete parameter valuation, and a property to evaluate. We cover parametric analysis in Sect. 7.

tool. Nevertheless, as QComp 2020 was presented at ISoLA 2020/2021, it became clear that more and more solution methods for advanced problems were being turned into tools of their own or integrated into existing stable tools such as PRISM [163] or STORM [125]. Therefore, the next edition of QComp that we present in this report, QComp 2023, shifts its focus towards these *frontiers in quantitative verification.*

Aims. The aims of QComp 2023 are (i) to describe advanced problems in quantitative verification for which analysis algorithms have more recently been developed and first tool support is appearing, (ii) to document the state of the art of this tool support, in terms of what is available today and what pieces are still missing, and (iii) to perform the first comparative tests of these tools where appropriate. The outcome of QComp 2023 is this competition report, which can serve as a guide to state-of-the-art tools for the domain expert faced with an advanced quantitative verification problem, as a historical reference for tool developers, and as a call to action pointing researchers to where better algorithms are still needed and tool developers to where "market opportunities" exist that can be filled with new tools.

Setup and Process. QComp 2023 is a more friendly competition than ever: It started with an open call for participation to the quantitative verification community in summer 2022. The interested participants then followed an iterative process of determining *categories* (i.e. advanced problem scenarios) of interest, which included identifying and contacting additional participants. Out of the group of all participants, we then established category coordinators who would lead the process needed to achieve the aims of the competition in their category. As the QComp categories covered all kinds of research and tooling maturity levels, part of the task of the category coordinators was to establish the scope and refine the concrete aims of their category. In categories where several sufficiently stable tools already exist, coordinators could choose to include a performance evaluation, while more cutting-edge categories would focus on a description of the category, available approaches, and prior experimental results if available. The category coordinators delivered the outcomes of their category to the overall QComp 2023 coordinator before summer 2023; over that summer, we integrated all contributions into this report.

In this distributed and flexible approach where the competition is divided into sub-groups that establish the actual aims of their own, QComp 2023 was modelled after the ARCH-COMP friendly competition on verifying continuous and hybrid systems (see cps-vo.org/group/ARCH/FriendlyCompetition), which has been running on this model successfully for seven editions as of today since 2017 [90], with its latest edition concluded just before QComp 2023 this summer.

2 Categories and Participants

As a friendly competition, QComp 2023 was open to all interested parties for suggesting, coordinating, and participating in categories related to quantitative verification. All participants of QComp 2023 are co-authors of this competition report. The competition as a whole was coordinated by A. Hartmanns. Before presenting the results of the individual categories in the remainder of this report, we give an overview of QComp 2023's ten categories with credits to the respective organisers and participants, and present the participating tools.

2.1 Categories

Infinite-state and population models (∞-*state*, Sect. 3): coordinated by Z. Zhang; participants: M. Češka, E. M. Hahn, J. Jeppson.

Long-run average rewards (*LRA*, Sect. 4): coordinated by K. Grover, J. Křetínský, and M. Weininger; participants: A. Hartmanns, T. Meggendorfer, and T. Quatmann.

Linear temporal logic (*LTL*, Sect. 5): coordinated by J. Křetínský and M. Weininger.

Multi-objective analysis (*multi-obj.*, Sect. 6): coordinated by T. Quatmann; participants: K. Grover, D. Parker, and M. Weininger.

Parametric Markov models (*parametric*, Sect. 7): coordinated by S. Junges.

Partially-observable MDPs (*POMDPs*, Sect. 8): coordinated by A. Bork; participants: R. Andriushchenko and D. Parker.

Rare events (*rare events*, Sect. 9): coordinated by C. E. Budde; participants: B. Israelsen, E. Ruijters, L. Taylor, M. Volk, and Z. Zhang.

Robust decision-making under uncertainty (*uncertainty*, Sect. 10): coordinated by N. Jansen; participants: D. Parker.

State space exploration (*exploration*, Sect. 11): coordinated by M. A. Köhl; participants: A. Hartmanns and T. Quatmann.

Stochastic games (*st. games*, Sect. 12): coordinated by D. Parker; participants: B. Könighofer, T. Meggendorfer, S. Pranger, and M. Weininger.

2.2 Participating Tools

Various tools ranging from research prototypes to mature toolsets are available today to tackle the problems covered by the different categories. In Table 1, we list which tools participated in which of the categories of QComp 2023. The meaning of "participate", however, can have a very different meaning in different categories; for example, the *parametric* category only names the four tools that support the analysis of parametric Markov models, while the *multi-obj.* category benchmarks its five participating tools and reports on their relative performance. Categories that include an experimental evaluation such as run-time benchmarking are indicated by a "Y" in the row labelled "*experiments*"; then row "*benchmarks*" states the number of benchmark instances considered in the experimental evaluation[3]. Participation of a tool in any category was voluntary and not automatic; in particular, if a tool does not participate in a certain

---

[3] More benchmarks may be *available* for the problems covered by a category, and category *parametric* has no performance evaluation but introduces a benchmark set.

Table 1. Tools participating in QComp 2023's different categories

	∞-state	LRA	LTL	multi-obj.	parametric	POMDPs	rare events	uncertainty	exploration	st. games
experiments	N	Y	N	Y	N	Y	Y	N	Y	Y
benchmarks	–	20	–	66	–	3	10	–	229	16
DFTRES							✓			
EPMC			✓	✓	✓					✓
FIG							✓			
INFAMY	✓									
MCSTA		✓							✓	
MODES							✓			
MOMBA									✓	
MULTIGAIN		✓	✓	✓						
PARAM					✓					
PAYNT						✓				
PET		✓								✓
PRISM			✓	✓	✓	✓		✓		
PRISM-GAMES		✓		✓						✓
RAGTIMER							✓			
SEQUAIA	✓									
STAMINA	✓									
STORM			✓	✓	✓	✓	✓		✓	
STORMDFTRES							✓			
TEMPEST		✓								✓

category, this does *not* imply absence of support for the advanced properties or model types that the category focusses on in the tool. To allow the individual category sections to focus on the specifics of their topic, we briefly introduce all 19 tools:

DFTRES [48], available at github.com/utwente-fmt/DFTRES, is a statistical model checker designed for repairable dynamic fault trees (DFTs [200]) specified in Galileo and more general CTMCs specified in JANI. It is written in Java and is portable to, at least, Linux, Windows, and macOS.

EPMC [91], available at github.com/iscas-tis/ePMC, is an extensible probabilistic model checking framework mostly written in Java. It is a successor of IscasMC [109].

FIG [43], available at git.cs.famaf.unc.edu.ar/dsg/fig, is a statistical model checker for transient and steady state reachability properties in CTMCs and input/output stochastic automata (IOSA) [71]. FIG is written in C++ and runs on Linux.

INFAMY [106], available at depend.cs.uni-saarland.de/tools/infamy, is a tool with the purpose of model checking formulae in continuous stochastic logic (CSL) [14,21] on infinite-state CTMC specified in a variant of the PRISM language by exploring the model up to a certain depth repeatedly. INFAMY can also handle certain reward properties.

MCSTA, available at modestchecker.net, is the explicit state model checker of the MODEST TOOLSET [113], a collection of tools for the modelling and analysis of stochastic timed and hybrid systems. Its core functionality is the disk-based explicit-state model checking of MDPs [114], MAs [51], PTAs [112], and stochastic timed automata [104]. The MODEST TOOLSET is mainly written in C# and runs on 64-bit Linux, macOS, and Windows systems. It supports the Modest [30,105] and JANI [44] input languages.

MODES [41], available at modestchecker.net, is the MODEST TOOLSET's statistical model checker. It supports the same input languages and platforms as MCSTA. It contains simulation engines specialised to different formalisms from DTMCs to stochastic hybrid automata with general probability distributions (SHA) [89], including support for non-linear continuous dynamics [186].

MOMBA [145], available at momba.dev, is a Python library centred around JANI with the goal of providing easy access to quantitative modelling capabilities.

MULTIGAIN [37] is an extension of PRISM for multiple long-run average rewards. MULTIGAIN 2.0 [23], available at zenodo.org/records/10610642, builds on MULTIGAIN, adding support for verification and strategy synthesis for LTL.

PARAM [107], available at depend.cs.uni-saarland.de/tools/param, was the first tool implementing verification algorithms for parametric Markov models.

PAYNT [9], available at github.com/randriu/synthesis, is a tool originally developed for the inductive synthesis of probabilistic programs. It aims at directly synthesising finite-state controllers for partially-observable MDPs.

PET [179] available at gitlab.lrz.de/i7/partial-exploration, is a model checker focusing on value iteration approaches augmented by partial exploration, based on [36] for reachability with subsequent extensions to mean payoff [12] and cores [148]. It is backed by tailored data structures and algorithms for this purpose, and implemented in Java.

PRISM [163], available at prismmodelchecker.org, is a widely-used probabilistic model checker supporting a large range of models and temporal logics. It is a user-friendly tool that comes with a cross-platform graphical user interface. PRISM is mostly written in Java, with some algorithms implemented in C.

PRISM-GAMES [158], available at prismmodelchecker.org/games, is an extension of PRISM focused on the verification of stochastic games.

RAGTIMER [129,211], available at github.com/fluentverification/ragtimer, is designed for chemical reaction networks (CRNs) modeled as CTMCs, combining guided stochastic simulation and commutability properties to compute lower-bound rare event probabilities from a partial state space.

SEQUAIA [54], available at sequaia.model.in.tum.de, offers two powerful engines for the quantitative analysis of population models given as chemical reaction networks via abstraction and simulation. Both build on an interval population abstraction of the underlying CTMC. SEQUAIA comes with a GUI, allowing for convenient modelling and tweaking the models as well as displaying the abstractions and analyses results for better explainability.

STAMINA [134,183,184,199], available at staminachecker.org, is an infinite-state PMC tool that iteratively explores a partial state space for a bounded or unbounded CTMC model. The CTMC transient analysis on the partial state space is delegated to PRISM's and STORM's PMC engines. STAMINA/PRISM implements the STAMINA 2.0 algorithm and interfaces with PRISM's Java API and uses PRISM for model parsing and checking. STAMINA/STORM is a reimplementation and extension of the STAMINA 2.0 algorithm using STORM.

STORM [125], available at stormchecker.org, is a general purpose, high-performance feature-rich probabilistic model checker built around a modular core with an emphasis on time and memory efficiency. Written in C++, STORM's modular design enables the utilization of different model checking engines catering to the characteristics of different models. Notably, STORM excels in efficient symbolic model checking through its dd engine leveraging binary decision diagrams (BDDs).

STORMDFTRES, available at gitlab.utwente.nl/fmt/fault-trees/storm-dft-res, implements multi-threaded Monte Carlo simulation for (non-repairable) DFTs given in either the Galileo or a custom format. STORMDFTRES builds on the STORM-DFT library [217] of STORM, which implements efficient state space generation for DFTs by exploiting e.g. irrelevant failures and symmetries.

TEMPEST [192] available at tempest-synthesis.org, is based on the STORM model checker, extending its feature set to turn-based stochastic games with a focus on synthesizing most-permissive strategies.

3   Infinite-State and Population Models

In many biochemical reaction and synthetic biology applications, very complex systems are studied and thus software tools become very advantageous and even indispensable for their understanding. For instance, the signalling pathways, chemical reaction networks, and genetic regulatory networks under study consist of many concurrent reactions running at very different speeds and probabilities, with species of both low and high copy numbers. This results in stiff systems suffering from stochasticity/multi-modality and state-space explosion, respectively [95,212], calling for dedicated analysis tools.

In order to analyse such systems, so-called *population models* are considered. A state of a population model is a tuple of integers, with the i-th component representing the copy number of the i-th species. Hence the state space is typically (countably) infinite. Transitions between states represent executing one reaction of the system. Given that the timing aspect is crucial and that the probability for a reaction to occur is (approximately) exponentially distributed as a function of real time, the model can be defined as a CTMC. This explicit model can be derived directly from a symbolic representations of the system as, say, a *chemical reaction network* (CRN): The rates of the CTMC can be computed from the rates of the CRN reactions and the copy numbers in each state using the mass action kinetics.[4] This transformation immediately enables the applicability of probabilistic model checkers for CTMC to biological systems. However, in order to make the analysis practical, the population structure has to be exploited in dedicated ways. In particular, one has to deal with the huge and in general infinite state spaces.

To handle such state spaces, various **reduction techniques** have been proposed that either truncate states of the underlying CTMC with insignificant probability [182] or leverage structural properties of the CTMC to aggregate/lump selected sets of states [1,16]. The *interval abstraction* of the species population is a widely used approach to mitigate the state-space explosion problem [223]. Alternatively, several hybrid models have been considered, such as treating only small-population species stochastically while using a deterministic semantics for large-population species [126], applying a moment-based description for medium/high-population species [121], or using the LNA approximation with an adaptive partitioning of the species according to leap conditions [53].

The investigated **properties** range from transient ("What is the (distribution over) states at time t?") to steady-state analysis (concerning the limiting distribution or LRA reward). The typical output of a tool for such a query is either a certain probability bound or an exact probability (or probability bound) of the predicate being true. Given the numeric character of the results and methods, approximate solutions are considered. Further, in contrast to verification, given that the systems are mostly neither safety-critical, nor completely modelled, it is typically acceptable to produce results without precise error bounds: often by simulation-based techniques [93] or aggressively practical, e.g. semi-quantitative [55], model-based approaches.

Table 2. Feature comparison of tools for population models

Tool	Platforms	Approach	Models	Syntax	Semantics
INFAMY	Linux	model checking + state truncation	CTMC	PRISM	CTMC
SEQUAIA	multi-platform	population abstraction + numerical, simulation	population models	GUI,dedicated	CTMC
STAMINA	Linux, macOS	model checking + state truncation	CTMC	PRISM	CTMC

---

[4] Consequently, more symbolic models such as stochastic Petri nets are hard to use since the rate of a transition for a particular reaction differs from state to state.

3.1   Tool Support and Benchmarks

The main technical characteristics of the available tools participating in this category of QComp 2023 are listed in Table 2.

INFAMY model-checks infinite-state CTMC specified in a variant of the PRISM language. It is capable of handling the time-bounded subclass of the logic CSL and certain reward properties. It explores the model up to a certain depth repeatedly while descending into the nested CSL formula. INFAMY provides different means for finding a stopping criterion for the state-space exploration. This is because there is a trade-off between when to stop and the memory needed to store the finite truncation of the state space.

SEQUAIA offers two engines, both building on a "population" abstraction of the underlying CTMC, abstracting concrete copy numbers to given intervals. The first one [54] computes an abstraction of the CTMC using *acceleration*, abstracting not only states and single transitions, but taking into account sequences of transitions. The resulting model is (i) small enough to *explain* the overall dynamics, and (ii) despite the induced imprecision, allows for a *semi-quantitative analysis*, computing not the exact probabilities of different behaviours, but their orders of magnitude, which is often sufficient in the biological applications. The engine thus features unprecedented scalability, analysing standard complex benchmarks within a fraction of a second, while it is precise enough to conclude on the qualitative behaviour of the system including rare behaviours and on rough estimates of the quantities (population sizes, times). The second engine provides a more precise quantitative analysis by uniquely *combining* the population abstraction with advanced simulation techniques [124]. It is based on a memorization technique that combines previously generated *segments* of runs defined over abstract states to generate new simulations more efficiently while preserving the original system dynamics and its diversity. It adapts online to identify the most important abstract states and thus utilizes the available memory efficiently. In combination with a novel fully automatic and adaptive hybrid simulation scheme, this speeds up the generation of trajectories and correctly predicts the transient behaviour of complex stochastic systems.

STAMINA iteratively explores a partial state space where a majority of the probability mass resides. It expands the state space *on the fly* based on the estimated state reachability probability, and truncates a search path when the estimate drops below a user-specified threshold. STAMINA then performs time-bounded transient PMC analysis by interfacing with PRISM or STORM. In this way, it computes a lower and upper bound, P_{min} and P_{max}, respectively, such that the actual probability of the CSL property under verification lies within $[P_{min}, P_{max}]$. The tightness of the probability window, $w = P_{max} - P_{min}$, is specified by the user, albeit with higher run-time for a smaller w. STAMINA can efficiently produce an accurate probability bound for CTMCs with an extremely large or infinite state space. It is not restricted to specific types of input models as long as they can be modelled as CTMCs using the PRISM modelling language. Examples include genetic regulatory networks [86,173], biochemical reaction systems [76,157], dynamic fault trees (DFTs) [217], and queuing network models [127,131]. STAMINA has been designed to support multiple exploration methods, and can be tailored to the model or property under verification. It has also been designed to be user-friendly and modular. Additionally, a graphical user-interface (written in Qt5) is under active development and will enhance user-experience and ease of use of STAMINA.

Benchmarks include the CTMC models in the QVBS, the PRISM Benchmark Suite [164], and the INFAMY case studies[5]. In addition, the Stochastic Model Case Studies repository [49][6] hosts a large collection of case studies focusing on biochemical systems with infinite state spaces. STAMINA has been evaluated on selected CTMC benchmarks from these benchmark suites, e.g. [134,199]. SEQUAIA has been evaluated on models describing challenging CRNs from the literature [54].

3.2   Outlook

Aside from the scalability limitation the tools are trying to mitigate, there are specific challenges in the analysis of biochemical and synthetic biological systems.

First, it is a very strong assumption that the correct model is available. Consequently, the analysis methods should be able to effectively handle various forms of **model uncertainty**, including unknown reaction rate parameters, unknown reactants or products, as well as unknown species bounds. The uncertainty can be modelled by various formalisms such as parametric or interval CTMC (see Sect. 10), for which the existing tools offer only very limited support.

Second, **concurrency** is fundamental to these systems, as their constituent chemical reactions are often simultaneously enabled. All enabled concurrent reactions may occur in a state but with very different probabilities, and their noisy operating environment can introduce extremely infrequent but potentially detrimental faults (see Sect. 9). Additionally, their regulatory nature and constituent reversible reactions can cause **cyclic behaviours** and often require long reaction execution sequences to reach a desirable state.

Finally, the verification tools should offer to the users (i.e. biologists) not only the verification result, but also an artifact in the form of a critical sub-system or a critical set of paths allowing the users to **interpret and explain** the results. While some rudimentary effort has been made, e.g. [55], this field is wide open.

4   Long-Run Average Rewards

Many frequently-studied classes of properties of probabilistic systems are based on *rewards*. A reward function assigns to every state (or action or state-action pair) a number modelling a cost (or a payoff) related to the single move. These rewards are accumulated over infinite paths in various manners. Popular ways

---

[5] https://depend.cs.uni-saarland.de/tools/infamy/casestudies/.
[6] https://github.com/fluentverification/CaseStudies_StochasticModelChecking.

are discounted, total, and average rewards [194]. While the *discounted reward* is heavily used in diverse applications ranging from economics to robotics, and is very easy to optimize, it essentially reflects a limited time horizon only. The *total reward* can reflect longer horizons better (e.g. unbounded reachability), yet not really the infinite-run behaviour. The *average reward* (also known as long-run average reward, limit-average reward, steady-state reward, or mean payoff) captures much more adequately the performance over an unknown or variable horizon (see e.g. [203]). Consequently, it is used to model e.g. performance properties, such as the average delay between requests and responses, the average rate of a particular event, etc. Considering the infinite horizon makes both classic and learning algorithms less efficient. The whole problem is thus more difficult, and also less studied in the context of AI or robotics. In contrast, it is significantly studied in formal verification where performance and dependability are critical and hard guarantees are desirable. Related to the average reward and reducible to it are constraints on the steady state of a system, which become more studied also in the context of AI; see e.g. [146,214]. The algorithms for long-run average (LRA) reward properties again span the whole spectrum of linear and dynamic programming, including value and strategy iteration, with the usual advantages and disadvantages. A specific case is the traditional steady-state analysis on (fully stochastic) Markov chains. There, solving a system of linear equations is sufficient, but for efficiency reasons often replaced by value iteration, too.

Table 3. Feature comparison of tools for average-reward properties

Tool	Objective	Model	Guarantees
MCSTA	ELRA	CTMC, MA	ε
MULTIGAIN	ELRA, SS	MDP	E-FP
PET	ELRA	DTMC, CTMC, MDP	ε
PRISM-GAMES	ELRA	TSG	none
STORM	ELRA, SS	DTMC, CTMC, MDP, MA	ε, E-RA
TEMPEST	ELRA	TSG	none

4.1 Algorithms and Tool Support

Table 3 gives an overview of tools supporting average-reward properties, differentiating the exact kind of supported objective (SS: steady-state or ELRA: expected long-run average reward), the supported models (where MA are Markov automata [79] and TSG are turn-based stochastic games, see Sect. 12), and the guarantees provided on the precision of the result (either none, ε-precise, E-FP: exact up to floating-point precision, or ERA: exact using rational arithmetic). We complement this high-level overview with short tool descriptions:

MCSTA supports [51] model-checking LRA reward properties in MA (and thus also in CTMC as a special case) using either a reduction to linear programming [101] or an ε-precise method based on value iteration [52].

MULTIGAIN implements a linear programming-based approach [35] for multi-objective steady-state and LRA reward objectives in MDP (see also Sect. 6).

PET focuses on partial-exploration techniques, for which it includes an extension to mean-payoff objectives [12].

PRISM-GAMES supports a wide range of zero-sum properties; for LRA rewards (and its multi-objective variant of *ratio objectives*), the stochastic games are required to be controllable multichains, i.e. the sets of states that can occur in any maximal end component must be almost surely reachable.

STORM can answer a wide range of queries for many different models, offering both value iteration-based approaches providing ε-precise results as well as linear programming-based algorithms using exact rational arithmetic. We highlight that STORM is the only tool able to handle negative rewards directly, while others require the rewards to be rescaled first (see Appendix A in the full version of [152] for the standard transformation).

TEMPEST implements mean-payoff analysis for TSGs on top of STORM, using value iteration with explicit state space representations.

4.2 Performance Comparison

Our performance comparison is only on MDPs, as this is the model that most tools support. Thus, we ran STORM (using the default value iteration which provides ε-guarantees), MULTIGAIN (using linear programming), and PET (using value iteration and partial exploration). We collected benchmarks from several sources [2,23,119]; Appendix G.1 of the full version of [2] contains descriptions of many of them. The experiments were conducted on a freshly installed Ubuntu virtual machine on top of an Intel i7-1165G7 CPU and 8 GB RAM. Each run had access to all eight cores available in the virtual machine and the tools were executed sequentially using a bash script, starting with STORM (on all benchmarks) and ending with PET. Table 4 shows for every benchmark some characteristics of the model and then the time in seconds taken by each tool (where MG 2.0 is MULTIGAIN 2.0) to compute the value; the best time is highlighted in bold. For all but one benchmark, STORM outperforms both MULTIGAIN and PET. The single exception is mer (4), where PET is slightly faster, leveraging the fact that only a very small part of the model has to be explored.

Data Availability. All model files used for the comparison, as well as the resulting log files, are available at DOI 10.5281/zenodo.8219191 [100].

4.3 Outlook

We pinpoint several streams of research currently pursued. First, the algorithms have been extended to **multiple average rewards** [35,61] and we refer to Sect. 6

Table 4. Performance comparison results of tools for average-reward properties

Model (Parameters)	# states	Value	STORM	MG 2.0	PET
busyRing	1,912	1.0	**0.04**	1.66	2.42
coin (N = 2, K = 5)	656	1.0	**0.01**	0.57	3.36
consensus (N = 4, K = 10)	104,576	1.0	**4.83**	189.95	4,505.29
csma (N = 2, K = 4)	4,958	1.0	**0.04**	1.52	4.49
cs_nfail	184	0.33	**0.02**	0.48	2.85
eajs (energy_capacity = 500)	93,228	3.64	**0.36**	11.02	168.80
eajs (energy_capacity = 1000)	193,728	3.64	**0.71**	36.61	345.28
firewire (deadline = 20, delay = 2)	2,862	0.0	**0.04**	1.15	6.22
ij (10)	1,023	1.0	**0.02**	0.83	3.61
investor	6,688	0.95	**0.07**	4.19	5.15
mer (3)	15,622	1.5	**0.81**	12.77	16.17
mer (4)	119,305	1.5	41.82	2,385.99	**41.52**
pacman (MAXSTEPS = 10)	6,852	0.78	**0.16**	3.41	4.71
pacman (MAXSTEPS = 15)	96,894	0.99	**2.95**	14.31	9.77
pnueli-zuck	2,701	1.0	**0.04**	1.31	3.91
rabin (3)	15,622	0.86	**0.23**	11.08	8.25
sensors (K = 5)	267	0.45	**0.01**	0.60	1.97
virus (3)	809	0.0	**0.02**	1.15	2.67
wlan (COL = 6, MAX_BACKOFF = 3)	284,446	0.0	**1.26**	9.04	24.82
zeroconf	1,088	0.0	**0.02**	0.70	1.79

for a discussion of the achievements and challenges. Second, some approaches handle **unknown models** that can only be simulated [2,153], or avoid their construction for efficiency reasons [148]. Finally, while value iteration is the prevailing solution approach, **guarantees on its precision** (stopping criteria) have only been given recently [12,152].

5   Linear Temporal Logic

The traditional analysis of MDPs, in particular in the context of operations research and performance optimization, is based on rewards. In domains such as AI, robotics, or economics, it is often the discounted reward; in other contexts, where the steady state or the long-run behaviour is more relevant, it is the average reward (see Sect. 4). However, in the context of verification, be it of hardware, software or cyber-physical systems, not only reachability but also more complex temporal properties are required [191]. While the analysis of branching-time properties typically boils down to reachability analysis, the analysis of linear-time properties is more complex. The most prominent formalism for capturing linear-time properties is the *linear temporal logic* (LTL) [191].

The standard way to analyse LTL properties is the automata-theoretic approach [213]: The formula is translated to an automaton and, subsequently, the product of the system and this automaton is analysed. While LTL properties occurring in verification of hardware or non-stochastic software tend to be very complex (see e.g. the LTL Store collection [149]), this is less pronounced for stochastic systems. One of the main reasons was the infeasibility of obtaining small automata apt for probabilistic model checking. Indeed, instead of nondeterministic Büchi automata (NBA), for which good translators had long been available, some degree of determinism is needed for MDP. Until recently, Rabin automata produced by determinisation were the default but hardly scalable solution. For instance, about 10 years ago, one fairness contract was translated by the then state-of-the-art methods within PRISM to an automaton with 4 states, while a conjunction of two yielded over ten thousand states, and a conjunction of three would not finish computing in a week [147].

In the decade since [147], alternative approaches started flourishing. They avoid the determinisation by direct translations [81] or by employing weaker forms of determinism, such as limit-determinism [108,205]. The resulting tools such as RABINIZER [151] or SPOT [77], or libraries such as OWL [150], now reach the same level of scalability as for nondeterministic automata, allowing for verification of more complex properties. Model checkers such as PRISM today (i) contain a pre-computed, built-in set of automata for the handful most used properties, and (ii) can link external translators to provide the state-of-the-art-sized automata via the Hanoi Omega-Automata (HOA) standard format [15]. Consequently, comparing the efficiency of model checkers themselves is not very relevant in this category; instead, we describe the main line of model checking algorithms, and discuss the limitations and additional features of the tools.

5.1 Algorithms and Tool Support

The standard algorithm (i) constructs the product of the system and the automaton (with a given acceptance condition), (ii) identifies the maximum accepting end components, and (iii) computes the reachability probability of their union. Step (ii) depends on the particular acceptance condition. The default since the inception of PRISM was the Rabin condition due to the better efficiency compared to Muller or parity. Since the appearance of the new translations, the algorithm has been extended to generalized Rabin [56] within PRISM and limit-determinism in the PRISM-based MOCHIBA [206] as well as in a lazy variant [108]. Further improvements on the sizes and types of the automata (e.g. good-for-MDP, Emerson-Lei) followed [82,110,135,174,181].

Several current tools can be applied to LTL and related specifications:

EPMC supports the verification of MDPs against LTL and PCTL*. It translates formulae to an NBA using SPOT and then applies the lazy approach [108] to compute its satisfaction probability.

MULTIGAIN 2.0 is an extension of PRISM capable of verification and strategy synthesis for LTL properties combined with long-run average rewards and steady-state constraints.

PRISM itself supports LTL and PCTL* model checking for MDPs. LTL model checking is done via deterministic ω-automata, typically (generalised) Rabin automata, simplified to Büchi or finite automata if appropriate. The translation uses a custom version of ltl2dstar, combined with a built-in automata library; alternatively, it can connect to external translators via the HOA format. For the subclass of (co)safe LTL, **PRISM** also supports cumulative expected reward until satisfaction properties.

STORM answers LTL, lexicographic multi-objective LTL [60], as well as PCTL* queries for MDPs. It uses deterministic ω-automata with general Boolean acceptance formulas obtained from SPOT.

5.2 Outlook

After a decade of research on alternative translations and automata types, the performance both in terms of runtime and of the near-minimality of the size of the automata have reached practical applicability. A few question remain open, such as whether the semantic notion of **good-for-MDP automata** allows us to produce yet smaller automata efficiently, compared to the syntactically defined acceptance conditions. However, the main focus should now move to **modelling and applications**. For LTL formulae, a decent amount of benchmarks is available. Many sets repeatedly used across different papers have been compiled in LTL Store [149]. However due to the earlier scalability problems in probabilistic LTL model checking, the number of probabilistic models coming together with more complex LTL specifications remains quite limited so far [119,164].

6 Multi-objective Analysis

System performance is commonly assessed with respect to multiple quantities such as the probability of a crash, the expected average energy consumption, or the expected time until task completion. System designers have to consider the interplay between these quantities: minimising the task completion time might require actions that increase the likelihood of a crash. *Multi-objective analysis* [62,83] reveals trade-offs between the considered quantities by showing which compromises are achievable. The system is given as a nondeterministic model $\mathcal{M}$ while the quantities are specified as a vector $\langle \varphi_1, \ldots, \varphi_\ell \rangle$ of $\ell \geq 2$ objectives. The objectives commonly refer to rewards attached to states or transitions of $\mathcal{M}$. An ℓ-dimensional point $\vec{p} = \langle p_1, \ldots, p_\ell \rangle \in \mathbb{R}^\ell$ is *achievable* if there exists a single policy for $\mathcal{M}$—i.e. a resolution of its nondeterminism—under which for all $i \in \{1, \ldots, \ell\}$ the value of objective

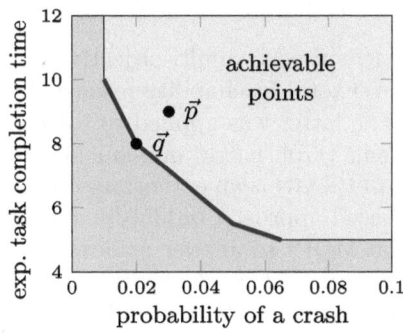

Table 5. Feature comparison of tools for multi-objective verification

	EPMC	PRISM	MULTIGAIN	STORM	PRISM-GAMES
models	MDP	MDP	MDP	MDP, MA	SG
objectives					
– reach. prob.	yes	yes	no	yes	qualitative
– total reward	yes	yes	no	yes	yes
– LRA reward	no	no	yes	yes	yes
– rew. bounded	no	steps	no	yes	no
– LTL prob.	yes	yes	yes	lexicographic	no
queries					
– achievability	yes	yes	yes	yes	yes
– numerical	yes	yes	yes	yes	no
– Pareto	no	$\ell = 2$	$\ell \leq 3$	yes	$\ell = 2$

φ_i is at least (or at most) p_i. Multi-objective analysis answers queries concerning the (set of) achievable points.

As an example, the green area in the figure above on the right shows the set of achievable points for $\ell = 2$ objectives as labelled on the axes. Point $\vec{p} = \langle 0.03, 9 \rangle$ is achievable but *dominated* by other achievable points; $\vec{q} = \langle 0.02, 8 \rangle$, for example, yields "better" values for both objectives. The blue line fragments indicate the set of undominated solutions—the *Pareto front*.

6.1   Algorithms and Tool Support

Table 5 compares quantitative verification tools in terms of their support for multi-objective analysis. We consider the supported kinds of models (where SG are stochastic games), objectives, and analysis queries. For the latter, we follow [88] and distinguish (i) achievability queries, asking whether a given point $\vec{p}$ is achievable, (ii) numerical queries asking for the optimal value for one objective while the others have to achieve a given $(\ell-1)$-dimensional point, and (iii) Pareto queries, asking for (an approximation of) the Pareto front.

We elaborate on the features of the individual tools:

EPMC supports multi-objective achievability and numerical queries for MDPs over total reachability reward objectives as well as objectives specified in LTL. The latter was applied to solve probabilistic preference-based planning problems [170]. EPMC implements the algorithm of [88] based on value iteration.

MULTIGAIN is an extension of PRISM that implements the linear programming-based approach of [35] for multiple steady-state and LRA reward objectives on MDPs to answer achievability, numerical, and Pareto queries—the latter for up to $\ell = 3$ objectives. A recent extension MULTIGAIN 2.0 [23] also incorporates the methods of [61,146] to add support for mixtures of steady-state, LRA reward, and LTL specifications.

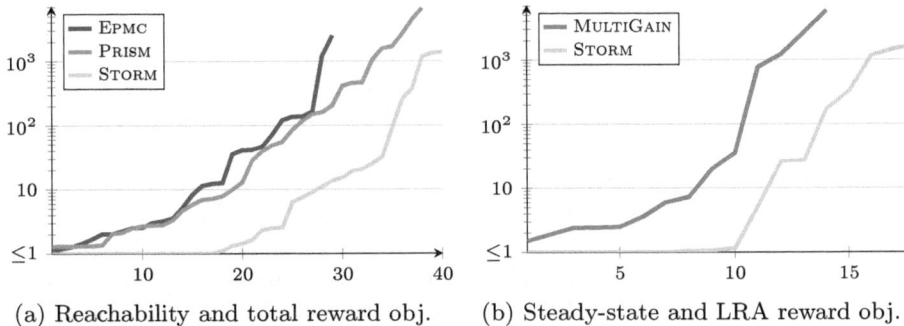

(a) Reachability and total reward obj.   (b) Steady-state and LRA reward obj.

Fig. 1. Performance comparison results of tools for multi-objective verification

PRISM answers achievability, numerical, and Pareto queries for MDPs over combinations of total reward, step-bounded reward, and LTL specification objectives. It implements methods based on value iteration [88] and on linear programming [87]. While the latter only works for achievability and numerical queries, the former can also be used to approximate Pareto fronts over up to $\ell = 2$ objectives. PRISM's graphical interface allows the user to conveniently examine the results.

PRISM-GAMES implements value iteration over convex sets [27] to analyze multiple total and LRA (ratio) reward objectives as well as almost-sure reachability constraints. PRISM-GAMES supports Pareto queries for $\ell = 2$ objectives and achievability queries for arbitrary Boolean combinations of objectives. An extension exists towards lexicographic queries for reachability objectives [59].

STORM handles achievability, numerical, and Pareto queries for MDPs and MA [196]. STORM implements the algorithm of [88] for total reachability reward objectives as well as extensions towards reward-bounded reachability objectives [115] and LRA reward objectives [197]. Furthermore, STORM supports multi-dimensional quantile queries [115], lexicographic LTL specifications [59], and multi-objective analysis under non-randomised policies with limited memory [74].

6.2   Performance Comparision

We empirically compare the performance of the tools for solving achievability queries on MDPs with (i) reachability and total reward objectives as well as (ii) steady-state and LRA reward objectives. We consider various benchmark models and objectives from the literature, e.g. [37,88,115,119,164]. To obtain challenging achievability queries, the queried points $\vec{p} = \langle p_1, \ldots, p_\ell \rangle$ have been obtained by roughly setting the threshold p_i for the ith objective to 90% of its optimal (single-objective) value. All experiments ran on an Intel Xeon Platinum 8160 CPU with 8 cores and 32 GB of RAM available. We measured the wall-clock runtimes of the tools and aborted executions after 1800 s.

Reachability and Total Reward Objectives. Our benchmark set contains 46 concrete queries over reachability probability and total reward objectives from 8 different model families. These queries are supported by EPMC, PRISM, and STORM, which all use the approach of [88] based on value iteration as their default method. On 12 queries, the tools reported inconsistent achievability results. We still include these problematic cases in our evaluation since identifying the *correct* solution is not trivial. The tool runtimes in seconds are summarised in Fig. 1 a. In this *quantile plot*, a point $\langle x, y \rangle$ on a line for a tool means that x out of the 46 queries *each* took at most y seconds to complete with this tool. We see that STORM is faster than both PRISM and EPMC. The competition among the latter two is tighter, with PRISM taking the lead.

Steady-State and LRA Reward Objectives. We consider 20 queries over steady-state and LRA reward objectives from 6 model families. We solve these queries using MULTIGAIN and STORM. All reported results were consistent for this set of experiments. Figure 1 b summarizes the runtime comparison (again as a quantile plot with runtimes in seconds). The implementation in STORM using the methods of [197] outperforms the linear-programming based approach of MULTIGAIN.

Data Availability. The benchmark models, scripts to reproduce the experiments, and our tool outputs are available at DOI 10.5281/zenodo.8063883 [195].

7   Parametric Markov Models

Classically, probabilistic model checking assumes that the probabilities on the transitions are fixed and precisely known. This assumption is often unrealistic: In various examples, such probabilities are approximations based on expert knowledge. In other applications, these probabilities reflect design decisions that can be freely made. *Parametric Markov models* replace constant probabilities by (polynomial) expressions over a fixed set of parameters X. A parametric Markov model and a valuation of its parameters induces a standard, parameter-free Markov model. The analysis of parametric Markov models was introduced almost 20 years ago [73, 168] while the tool PARAM brought first tool support more than 10 years ago [107]. By now, the model checkers STORM, PRISM and EPMC have support for parametric models.

Over the last decade, there have been various algorithmic advances that answer a variety of *different* queries about a parametric model [132]. The accompanying algorithms have been implemented in various tools and prototypes and all make different assumptions. Furthermore, not every benchmark is well-suited to motivate a particular query. This situation harms further adoption. In the spirit of QComp, we provide a unified and cleaned-up reference implementation on top of the probabilistic model checker STORM, and an annotated benchmark set for various parametric verification queries on parametric DTMCS (pDTMCs) and parametric MDPs (pMDPs).

7.1   Queries and Algorithms

We formulate the key verification tasks for parametric Markov chains. For conciseness, we assume that we are interested in computing the expected reward until reaching a target state, which generalizes computing reachability probabilities. For pMDPs, we assume that we consider the maximal expected reward. The key queries we identify are as follows:

Feasibility: Find parameter values such that the induced expected reward is above a threshold. The state-of-the-art methods rely on guess-and-verify and guess using sampling [64], gradient descent [122], and convex optimisation [67]; the former methods are fastest with few parameters and the latter are fastest with a larger number of parameters. For pMDPs, the quantification order is first over the parameters and then over the schedulers, i.e. the scheduler may depend on the parameter value chosen. This contrasts with *robust* schedulers that do not allow this [10, 219].

Verification: Show that no parameter values exist such that the induced expected reward is above a threshold. This problem is the dual to feasibility queries, but the universal quantification makes it harder to solve. The state-of-the-art approach employs an abstraction-refinement loop [138] using interval Markov chains and combines this with a graph-based analysis to determine monotonicity of the parameters [208].

Solution function computation: Compute a function that maps parameters to the induced expected reward in the corresponding Markov model. While various dedicated methods exist [85, 107, 138], linear equation solving over the field of rational functions performs great overall [138]. Theoretically, a one-step fraction-free method prevents intermediate blowups [22]. For pMDPs, the shapes of these functions are typically prohibitive.

A formal treatment is provided in [136]. Various other queries have been discussed in the literature. They aim to partition the parameter space [138], repair a model with the best parameter values [26], quickly sample parametric Markov models [92], or check whether the derivative is (globally) positive [207]. Others assume a distribution over parameter values [18, 34].

Practicalities. Typically, approaches limit the type of parameter valuations that are considered; graph-preserving valuations require that the underlying graph does not change. While this does not change the complexity of e.g. feasibility [219], it means that the solution function for pDTMCs is a (continuous) rational function and simplifies preprocessing. Likewise, most approaches assume that all pDTMCs are *simple(x)* [136], which (roughly) means that the transition probabilities are given by affine functions that syntactically sum to one. While it is theoretically relevant to allow real-valued parameter valuations, tools typically restrict themselves to rational (or floating-point) number representations.

7.2   Benchmark Collection

We provide a benchmark collection with 12 benchmark families at github.com/sjunges/parametric-Markov-models (with the models also archived at DOI 10.5281/zenodo.10646479 [137]). This benchmark collection includes reference invocations for STORM. The collection includes parametric versions of classical benchmarks [107,119,164] as well as benchmarks based on the usage of parametric verification in the analysis of hierarchical MDPs [140] and from the sensitivity analysis of Bayesian networks [202].

7.3   Outlook

We believe that the engineering behind many algorithms is still naive and that there is great potential for **algorithmic improvements**. In particular, the verification algorithms lack severely behind in their scalability, and despite being built on top of probabilistic model checking engines, most algorithms have only been implemented for expected rewards and reachability probabilities. More fundamentally, the **synthesis of robust policies in pMDPs** is an open challenge. A next iteration of QComp could also include parametric CTMCs [38], parametric PTAs [118], or structural parameters [9].

8   Partially-Observable MDPs

A major shortcoming of the classic MDP framework is the assumption that decisions can be made based on *complete* state information. In many domains where reasoning about uncertainty is necessary, this assumption is not realistic. For example, information about the current state of an autonomous vehicle is inherently incomplete as it perceives its environment through imperfect sensors.

Partially Observable MDPs (POMDPs) extend MDPs with the notion of *partial observability*. Nondeterminism is resolved not based on the complete state, but on the observable information available to the decision procedure. As such, policies for POMDPs are required to be *observation-based*, i.e. decisions are based on the observations and their history. We consider reachability objectives in POMDPs, i.e. we are interested in the minimal or maximal *reachability probability* of certain states in the system or, alternatively, in the minimal and maximal expected total reward until reaching a set of states. In contrast to fully observable MDPs, where optimal policies for such objectives that are memoryless always exist, in POMDPs memory is crucial even for sub-optimal policies.

While POMDPs are widely used for planning in domains like artificial intelligence [201], many verification and synthesis problems have proven to be undecidable. For example, even determining if the reachability probability of a set of states in a POMDP exceeds a threshold is undecidable [172]. Tool support for POMDPs exists in the AI community [78,155], however, these tools focus on *discounted* objectives, often over a finite horizon [204]. In recent years, efforts took place to extend the tool support for the verification setting of infinite-horizon objectives *without* discounting, also called *indefinite-horizon* objectives. Due to the undecidability of key problems in this setting, the applied methods focus on approximating values and synthesising *good* (sub-optimal) policies with respect to the objectives.

8.1   Algorithms and Tool Support

We showcase three tools from the formal methods community that deal with verification and policy synthesis for POMDPs.

PRISM includes support for POMDPs as well as a partially observable variant of PTA. It solves probabilistic reachability or expected cumulative reward queries using the model checking algorithm of [188], which implements a grid-based approach [171, 222] for computing an over-approximation of the objective value on an abstraction of the infinite, fully observable *belief MDP* of the POMDP using only a finite number of beliefs. The resulting policy is then solved to yield an under-approximation which, together, provides lower and upper bounds on the objective value for the POMDP. If the bounds are not tight enough, the approximation can be refined by increasing the grid resolution. The implementation builds upon PRISM's Java-based explicit-state engine. The tool then allows the resulting policy to be visualised or simulated in its graphical user interface.

STORM has support for POMDPs that is actively in development. In contrast to PRISM, over- and under-approximations can be computed independently of each other. For over-approximations, STORM implements an improved version of the grid-based approach from [171]: it allows for variable grid resolutions for different observations and on-the-fly generation of the belief grid [32]. For under-approximations, STORM uses *belief unfolding with cut-offs*: the belief MDP is unfolded starting in the initial belief. After a fixed number of beliefs has been unfolded, the objective value for all beliefs that have not yet been fully expanded are approximated. This approximation is based on values computed on the POMDP itself using *some* observation-based policy [33]. These values can be computed heuristically by STORM or provided externally. The abstract MDPs are then checked using STORM's MDP model checking core. In addition to the quantitative properties considered here, STORM also supports the verification of *qualitative* properties on POMDPs [139].

PAYNT was originally developed for the inductive synthesis of probabilistic programs. In contrast to the model checkers described above, it aims at directly synthesising *finite-state controllers* (FSCs) for POMDPs. An FSC is a Mealy machine that compactly represents a finite-memory policy. To find the best FSC within a given design space of controllers, an MDP abstraction is used, which encodes every possible decision and memory update a policy can make. The resulting process is an over-approximation in the sense that it can simulate every FSC in the design space and switch between FSCs mid-execution. Model checking the MDP yields the best policy within the design space and, if the policy is not observation-based, a refinement takes place. Additionally,

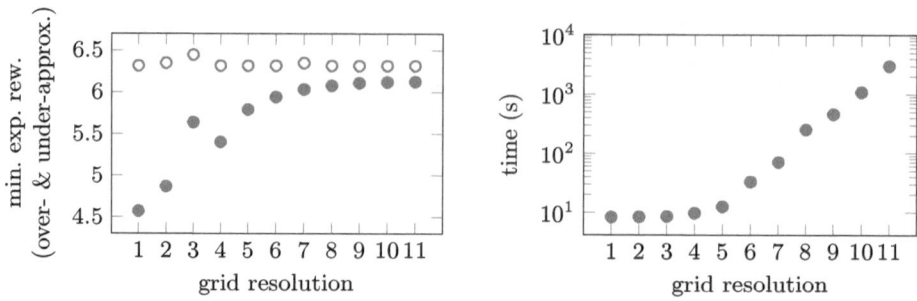

Fig. 2. PRISM's over- and under-approximations for *grid* and computation times

the design space can be pruned by generating counter-examples [8]. For computations on the MDP abstraction and assessing the quality of a synthesised FSC, PAYNT internally uses STORM. As PAYNT synthesises a policy, it can only provide under-approximations of the objective values.

8.2   Performance Comparison

We empirically evaluate the tools described above. As the different approaches and features of the tools make a direct comparison between them misleading, we consider the tools separately on different benchmarks. All three tools accept as input descriptions of POMDPs in the PRISM format. Furthermore, STORM and PAYNT allow inputs in the explicit DRN format.

From the repository of POMDP benchmarks from the literature [8,33,188] at github.com/moves-rwth/pomdp-collection, we select one benchmark for each tool to showcase some of its capabilities. All experiments ran on an Intel Xeon Platinum 8160 CPU using 2 threads, 32 GB of RAM, and a time limit of 1 h (measured in wall time). All tools are called using default configurations and options except for the input parameter we evaluate. The short evaluation presented here is not representative of the full capabilities of the tools. Tweaking the configurations used for running the tools typically leads to improvements in the results obtained.

PRISM. We consider the *grid* benchmark, instantiated with length 4 and slipping probability 0.3, for our evaluation of PRISM. The objective here is to minimise an expected reward. We evaluate PRISM with respect to changes in the resolution of the belief grid considered for the over-approximation. Our results are depicted in Fig. 2. The grid-based over-approximation—in the case of minimisation a lower bound—is depicted with solid dots while the under-approximation is depicted with circles. The experiments clearly show the impact of increasing the grid resolution. Generally, the higher the resolution, the better the computed bounds, for the over-approximation as well as the under-approximation which is only indirectly effected by the grid resolution. However, this improvement comes at the cost of greatly increased runtimes. The outlier at resolution 3 also shows that

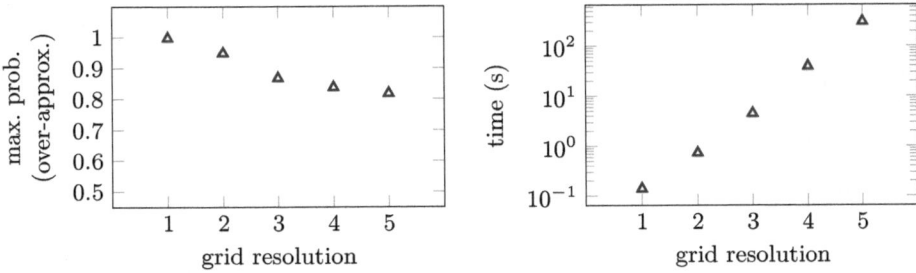

Fig. 3. STORM's over-approximations for *refuel* and computation times

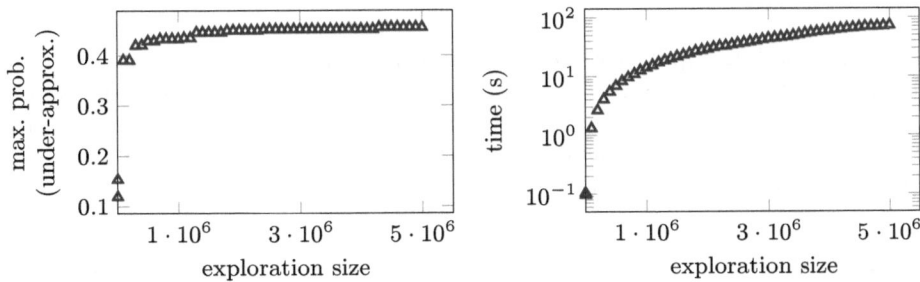

Fig. 4. STORM's under-approximations for *refuel* and computation times

a good choice of resolution may lead to a tighter approximation of the actual belief space even for rather small resolutions. For resolutions greater than 11, the tool either times out or runs out of memory.

STORM. We evaluate STORM on the *refuel* benchmark instantiated with length 10, computing the maximal probability. We study the over-approximation with respect to changes in the grid resolution and the under-approximation with respect to the size of the unfolding of the belief MDP. Figures 3 and 4 depict the respective results. Like for the over-approximation in PRISM, we observe that an increase in resolution for the grid-based approach leads to tighter bounds on the optimal objective value, with an impact on the runtime. For higher resolutions than 5, STORM ran out of memory on the benchmark. For the under-approximation, we consider unfoldings with increasingly larger state spaces. For larger unfoldings than depicted, we did not achieve better values. We see that with increasing unfolding depth, a tighter bound is achieved. This effect is particularly pronounced in the range of smaller unfolding sizes. With a linear increase in the unfolding size, runtimes appear to scale proportionally.

PAYNT. For evaluating PAYNT, we select the *grid-avoid* benchmark, instantiated with length 4 and slipping probability 0.1, where the objective is to maximise the probability to reach a target. While a key feature of PAYNT is its ability to dynamically increase the size of the considered FSC during computation, we focus on its functionality to compute FSCs of a given size. Thus, we vary

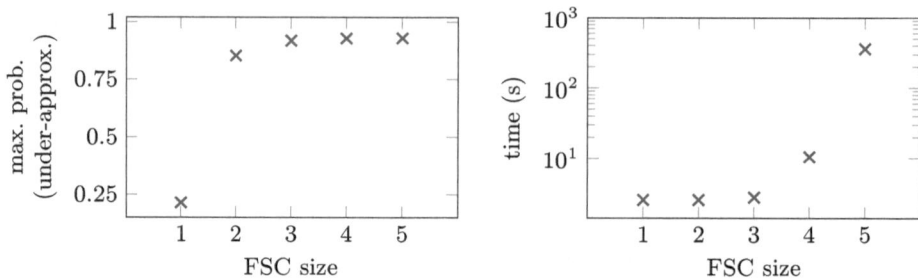

Fig. 5. Max. prob. achieved and time to compute the FSC for *grid-avoid* with PAYNT

the corresponding input parameter. Our results in Fig. 5 show that PAYNT is able to obtain FSCs using *some* memory very quickly while FSC quality greatly improves when memory is considered. With increasing FSC size, however, the effect is far less pronounced while runtimes increase drastically. For all FSC sizes greater than 5, the tool timed out in our experiment.

Data Availability. The benchmark models used in and the tool outputs generated by our experiments are available at DOI 10.5281/zenodo.8215337 [31].

9   Rare Events

In stochastic systems, *rare events* (RE) stand for measurable events with a positive but very low probability of occurrence. A typical example is the failure probability of highly-reliable systems that can be from 10^{-3} to 10^{-20} or lower.

Formal methods tools can encounter RE in quantitative computations of PRCTL- or CSL-like queries on any model with probabilities. Queries on general stochastic models—i.e. models in continuous-time with residence times or transition firings governed by arbitrary probability distributions—are often estimated via Monte Carlo simulation. This is known as *statistical model checking* and is hindered by RE: since the states of interest are seldom visited, either the number of simulation runs required and thus the runtime grows to impractical values, or the statistical estimations become imprecise or even incorrect. The field of *rare event simulation* (RES) has developed to tackle this problem, and can be divided in two main approaches: *importance sampling* (IS [123]) and *importance splitting* (ISPLIT [141]).

In contrast, exhaustive state-space exploration approaches such as (probabilistic) model checking are not hindered by the rarity of an event, which makes them very attractive to solve RE property queries. However, model checking struggles with the state-space explosion problem, which can be portrayed as the counterpart of the runtime explosion problem faced by simulation analyses. Additionally, no scalable exhaustive methods are available for models with general distributions. Thus, the challenge here—in the Markovian case—is how to reduce the model size without compromising the correctness or accuracy of the final estimate.

We compare tools that implement RES and model checking to estimate quantitative RE queries in formal stochastic models. Besides defining the scope and capabilities of each tool, we showcase their computation of RE queries in six models with Markovian and arbitrary probability distributions.

Table 6. Feature comparison of tools for rare event estimation

Tool		Approach to RE		Models		Semantic formalism
Name	OS	Type	Subtypes	Types	Syntax	
DFTRES	Linux, macOS, Windows	RES: IS	Path-ZVA, forcing	DFT, RFT	JANI, Galileo	CTMC, MA (subset)
FIG	Linux	RES: ISPLIT	RESTART-P$_j$, fixed effort	IOSA, DFT, RFT	IOSA, JANI, Galileo, Kepler	CTMC, IOSA
MODES	Linux, macOS, Windows	RES: ISPLIT	RESTART-P$_j$, fixed effort	any	Modest, JANI	DTMC, CTMC, SHA
RAGTIMER	Linux	probabilistic model checking	partial exploration	CRN	RAGTIMER	CTMC
ST.DFTRES	Linux, macOS	RES: ISPLIT	RESTART	DFT	Galileo	CTMC

9.1 Algorithms and Tool Support

Table 6 summarises the characteristics of tools in formal methods that can estimate rare events. Naturally, the list is not exhaustive: see e.g. earlier works [24,133,178]. In QComp 2023, we compare the following tools for rare event scenarios, of which all except RAGTIMER are statistical model checkers:

DFTRES is designed to estimate transient and steady-state properties of repairable DFTs specified in Galileo, and can also be applied to more general CTMCs and some MAs (that reduce to CTMCs) specified in JANI. It automatically applies IS RES, through the Path-ZVA algorithm [198] and, for transient properties, using forcing [185]. These techniques are particularly applicable for models in which the target event can be reached in a relatively small number of low-probability steps. The algorithms can be applied with no user adjustments, however manual tweaking can improve performance on specific models.

FIG estimates transient and steady state reachability properties in CTMCs and IOSA. It can parse the IOSA and JANI syntax for general models, and Galileo and Kepler for repairable DFTs. FIG automates the use of ISPLIT RES by deriving the importance function from the model and property query [42]. This simplifies the user input to choosing a thresholds-selection algorithm (sequential Monte Carlo or expected success [40]), a simulation run type (RESTART, RESTART-P$_j$, or fixed effort), and termination criteria (e.g. by runtime length). There are no theoretical proofs—e.g. of asymptotic efficiency—on the convergence time of the algorithms used by the tool on general models.

MODES implements ISPLIT to tackle RES with manually-specified or automatically-derived importance functions much like FIG, including support for the same run types and thresholds-selection algorithms [40, 45].

RAGTIMER uses guided stochastic simulation and commutability properties to build a partial state space and acquire a lower-bound for a rare-event probability in chemical reaction networks (CRNs) modeled as CTMCs. It uses reaction information to create a dependency graph, which can demonstrate unreachability. If a property is reachable, it constructs a probability-agnostic model for compositional testing in the IVy tool [177] and uses stochastic simulation to generate a large number of counterexample traces. It then expands these traces and discovers parallel traces by firing commutable reactions and cycles from every state along a trace. The resulting partial state space is passed explicitly to a probabilistic model checker to obtain a lower-bound on the probability of interest. In preliminary testing, RAGTIMER finds or approaches the true probability of rare-event properties in several CRN models.

STORMDFTRES analyses time-bounded reachability properties on (non-repairable) DFTs in either the Galileo or a custom STORMDFTRES format represented as CTMCs. It aims to perform importance splitting for RE following the ideas of FIG and using the importance functions for DFTs presented in [43].

9.2 Performance Comparison

We demonstrate the capabilities of the tools to compute various PRCTL- and CSL-like property queries on a series of CTMC and SA models. The models used for experimentation are summarised in Table 7: there are six Markovian (CTMCs) and one non-Markovian (SA) models, the latter with hyperexponential and Erlang distributions. All models are provided in the syntax of the tool that specialises in it, and which introduced it to this comparison. They have also been translated to JANI, for the model exchange across tools that enabled this comparison. The SA model is an exception: it is provided in the IOSA and MODEST syntaxes alone (for the FIG and MODES tools), since it has committed actions that are currently unsupported in JANI.

One or more rare-event properties are given per model: We used the tools to estimate them, showing the results in Table 8. Per model and property we had the tools run for 1, 5, 10, and 30 min (indicated in column ☻) in the TACAS VM [84]. Each tool could use a default run (minimal configuration) or custom commands for that model-property combination. Table 8 reports only one of those values: when the difference between them is below 15% we report the default run; else, we report the one closer to the exact property value [47]. In the table, an empty

Table 7. Models used in the comparison of tools for rare event estimation

Name	Type	Family	Description	Properties
forked-cycle-tandem-queue	CT MC	queueing system	*three queues*: arrivals to Q1; probabilistic output to Q1, Q2, Q3; study overflows in Q2. *(previously unpublished)*	φ_1: $\mathtt{P}_{=?}[\,\mathtt{q2} > 0 \ \mathtt{U} \ \mathtt{q2} \geqslant 6\,]$ φ_2: $\mathtt{P}_{=?}[\,\mathtt{q2} > 0 \ \mathtt{U}_{\leqslant 555} \ \mathtt{q2} \geqslant 6\,]$ φ_3: $\mathtt{P}_{=?}[\,\mathtt{F}_{\leqslant 555} \ (\mathtt{q2} \geqslant 6)\,]$ $\tilde{\varphi}_4$: $\mathtt{S}_{=?}[\,\mathtt{q2} \geqslant 6\,]$
7nodes-network	SA	queueing system	*non-Jackson 7 queues*: arrivals to all queues; near-full probabilistic interconnection; study overflow in Q7. [215]	$\tilde{\varphi}_5$: $\mathtt{S}_{=?}[\,\mathtt{n7} \geqslant 30\,]$
2react	CT MC	chemical reaction network	*single species production-degradation*: simple 2-reaction system with one shortest trace. [69]	φ_6: $\mathtt{P}_{=?}[\,\mathtt{F}_{\leqslant 100} \ (\mathtt{s2} \geqslant 80)\,]$
6react	CT MC	chemical reaction network	*enzymatic futile cycle*: 6-reaction system with large state space, cyclic behavior, and one shortest trace. [157]	φ_7: $\mathtt{P}_{=?}[\,\mathtt{F}_{\leqslant 100} \ (\mathtt{s5} = 25)\,]$
8react	CT MC	chemical reaction network	*modified yeast polarization*: concurrent 8-reaction system with cyclic behavior and many shortest traces. [76]	φ_8: $\mathtt{P}_{=?}[\,\mathtt{F}_{\leqslant 20} \ (\mathtt{G\_bg} \geqslant 50)\,]$
HECS	CT MC	dynamic fault tree	*hypothetical example computer system*: standard DFT benchmark. [216]	φ_9 : $\mathtt{P}_{=?}[\,\mathtt{F}_{\leqslant 1} \ \mathtt{TLE}\,]$ *"unreliability @ 1"*
MAS	CT MC	dynamic fault tree	*mission avionics system*: highly redundant safety-critical system with hard- and software components. [189]	φ_{10}: $\mathtt{P}_{=?}[\,\mathtt{F}_{\leqslant 1} \ \mathtt{TLE}\,]$ *"unreliability @ 1"*

cell indicates no support for that property/model. Values produced by a custom command are marked with a superscript star*. The values reported are either 95% confidence intervals ($p \pm \delta$), sound lower bounds ($\geqslant p$), failures ($\varnothing$), or omitted computations ("). The latter applies to e.g. model checkers like RAG-TIMER that use one runtime since longer runs are seldom beneficial. In general, smaller confidence intervals and results closer to the true value (indicated in the second column under heading "Prop." as either a statistical approximation $\approx p$ or truncated exact value p, obtained from reference material or computed with higher resources, viz. more memory, runtime, and cpu power) are better.

We note that the default ("sound") run of DFTRES can run longer or shorter than the hard time limit, and its renewal-theory implementation cannot compute $\tilde{\varphi}_4$ on that JANI model; also, FIG and MODES used crude Monte Carlo (not RES) to analyse the DFTs because no useful importance function could be derived when the dominant failures have short traces; and RAGTIMER used one runtime per property, since longer runs are seldom beneficial for model checkers.

Data Availability. We provide an artifact allowing a full experimental reproduction at DOI 10.6084/m9.figshare.23818395 [47].

9.3   Outlook

We see the need for further research to unify rare event approaches in the formal tools community, e.g. to allow **automatic identification of the algorithm to use**. A concrete example is the high performance of DFTRES (using IS) to analyse the DFTs in contrast to its comparatively low performance for properties in queueing systems. This in contrast to FIG and MODES, which (using ISPLIT)

Table 8. Performance comparison results of tools for rare event estimation

Prop.	ref	☉	DFTRES	FIG	MODES	RAGTIMER	STORMDFTRES
φ_1	9.23E-10	1	9.2E-10 ± 3E-13	9.0E-10 ± 9E-11*	9.4E-10 ± 6E-11*		
		5	9.2E-10 ± 6E-14	9.5E-10 ± 4E-11*	9.2E-10 ± 4E-11*		
		10	9.2E-10 ± 4E-14	9.0E-10 ± 3E-11*	9.2E-10 ± 2E-11*		
		30	9.2E-10 ± 2E-14	9.3E-10 ± 2E-11*	9.1E-10 ± 8E-11*		
φ_2	9.23E-10	1	9.2E-10 ± 5E-13	9.4E-10 ± 1E-10*	9.3E-10 ± 3E-10*		
		5	9.2E-10 ± 8E-14	9.2E-10 ± 5E-11*	9.1E-10 ± 3E-11*		
		10	9.2E-10 ± 6E-14	9.3E-10 ± 3E-11*	9.2E-10 ± 2E-11*		
		30	9.2E-10 ± 3E-14	9.2E-10 ± 2E-11*	9.3E-10 ± 1E-11*		
φ_3	9.00E-08	1	9.4E-09 ± 3E-09*	8.7E-08 ± 7E-08*	8.1E-08 ± 2E-08*		
		5	8.6E-09 ± 1E-09	7.6E-08 ± 3E-08*	9.2E-08 ± 6E-09*		
		10	1.1E-08 ± 4E-09	8.3E-08 ± 3E-08*	9.0E-08 ± 4E-09*		
		30	1.3E-08 ± 5E-09	9.1E-08 ± 2E-08*	9.1E-08 ± 3E-09*		
$\bar{\varphi}_4$	5.64E-11	1	∅	6.2E-11 ± 2E-11*			
		5	∅	6.0E-11 ± 6E-12*			
		10	∅	5.8E-11 ± 4E-12*			
		30	∅	5.5E-11 ± 2E-12*			
$\bar{\varphi}_5$	7.57E-15	1		7.0E-15 ± 5E-15*	7.1E-15 ± 2E-15*		
		5		8.3E-15 ± 4E-15	7.7E-15 ± 1E-15*		
		10		7.2E-15 ± 2E-15	7.7E-15 ± 6E-16*		
		30		8.8E-15 ± 3E-15	8.3E-15 ± 4E-16*		
φ_6	3.06E-07≈	1			2.9E-07 ± 1E-08*	⩾ 3.0E-07	
		5			3.0E-07 ± 7E-09*	"	
		10			3.0E-07 ± 5E-09*	"	
		30			3.0E-07 ± 3E-09*	"	
φ_7	1.70E-07	1			1.7E-07 ± 4E-08*	⩾ 2.8E-18*	
		5			1.8E-07 ± 2E-08*	"	
		10			1.7E-07 ± 1E-08*	"	
		30			1.8E-07 ± 8E-09*	"	
φ_8	1.20E-06	1			1.5E-06 ± 3E-07*	∅	
		5			1.5E-06 ± 1E-07*	⩾ 2.3E-28	
		10			1.6E-06 ± 9E-08*	"	
		30			1.7E-06 ± 5E-08*	"	
φ_9	2.20E-04≈	1	2.2E-04 ± 6E-06	2.3E-04 ± 3E-05	2.1E-04 ± 3E-05		2.2E-04 ± 2E-05
		5	2.2E-04 ± 5E-07	2.2E-04 ± 1E-05	2.2E-04 ± 1E-05		2.2E-04 ± 7E-06
		10	2.2E-04 ± 2E-07	2.2E-04 ± 1E-05	2.2E-04 ± 1E-05		2.2E-04 ± 5E-06
		30	2.2E-04 ± 1E-07	2.2E-04 ± 5E-06	2.2E-04 ± 5E-06		2.2E-04 ± 3E-06
φ_{10}	1.00E-05	1	1.1E-05 ± 8E-06	8.1E-06 ± 1E-06	6.7E-06 ± 3E-06		1.4E-05 ± 5E-06
		5	1.0E-05 ± 2E-06*	7.3E-06 ± 5E-06	1.2E-05 ± 2E-06		1.0E-05 ± 2E-06
		10	9.7E-06 ± 2E-06	1.0E-05 ± 5E-06	1.1E-05 ± 1E-06		9.8E-05 ± 1E-06
		30	1.0E-05 ± 1E-06	8.9E-06 ± 2E-06	9.7E-06 ± 7E-07		1.1E-05 ± 8E-07

performed well for the latter, but found crude Monte Carlo to be their best approach for the DFTs.

10   Robust Decision-Making Under Uncertainty

In recent years, there has been a strong push to combine the areas of formal verification—in particular model checking—and artificial intelligence (AI). A specific area that is native to both of those areas is *decision-making under uncertainty* [143]. The level and type of uncertainty affect the capabilities of AI systems to make intelligent decisions. The core problem is to provide a guarantee that an AI system, operating under uncertainty, adheres to some formally specified constraint, e.g. given as a temporal logic specification (see Sect. 5). State-of-the-art approaches use models, in particular MDPs, to capture sequential decision-making problems for agents operating in uncertain environments. Moreover, sensor limitations may lead to partial observability of the system's current state, giving rise to POMDPs (see Sect. 8). MDPs augmented with a model of adversarial behaviour are stochastic games (SGs, see Sect. 12) and their partially observable counterpart is a partially-observable SG (POSG).

The likelihood of uncertain events, such as a message loss in communication channels or specific responses by human operators, may only be an estimate from data. The models mentioned above capture uncertainty in the form of precise probabilities—either in their transition dynamics or in their observation models. However, such *point estimates* of probabilities from data carry the risk of statistical errors. Moreover, the optimal policies for agents are usually highly sensitive to small perturbations in transition probabilities, leading to suboptimal outcomes such as a deterioration in performance [96,175].

Uncertainty models, sometimes also referred to as *robust models*, remove this assumption by incorporating uncertainty sets of probabilities. In the literature, uncertain MDPs use, for example, *probability intervals* or *likelihood functions* [187,218]. Similar extensions exist for uncertain POMDPs, where uncertainty may also affect the observation model [39,50,68,130,209]. To the best of our knowledge, there are no results on uncertain POSGs. Figure 6 shows a family of *uncertainty models*, capturing different types of uncertainty and their relation to each other. The three different types of arrows indicate the addition of (1) adversarial behaviour, (2) uncertainty on probability distributions, and (3) partial observability from one model to another. In the figure, adversarial behaviour increases from left to right. The left and right columns are partially observable models. Finally, the bottom row shows models that (in addition to probabilistic and adversarial behaviour) account for uncertainty in probability distributions. For an overview, we refer the interested reader to e.g. [17].

10.1   Algorithms and Tool Support

PRISM [163], a widely-used probabilistic model checker, provides support for two common classes of uncertainty models: *interval discrete-time Markov chains*

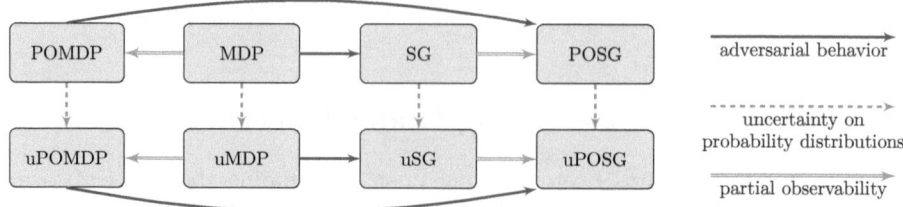

Fig. 6. A family of closely related uncertainty models

(IDTMCs) and *interval Markov decision processes* (IMDPs). These are specified to the tool with a simple extension of the PRISM modelling language where the probabilities attached to variable updates within a guarded command are optionally provided as intervals, for example (In this example, *move* labels the transitions induced by the command, $loc = 1$ is the guard that determines when the command is enabled, and each of the two branches to the right of $\rightarrow$ has an interval of probabilities and a set of assignments.)

$$[move]\ loc = 1\ \rightarrow\ [0.85, 0.95]\colon (loc' = 2) + [0.05, 0.15]\colon (loc' = 1);$$

and, as usual, can be given as expressions in terms of variables and parameters:

$$[send]\ s = 1\ \rightarrow\ [p_{fail} - \varepsilon, p_{fail} + \varepsilon]\colon (s' = 1) + [(1 - p_{fail}) - \varepsilon, (1 - p_{fail}) + \varepsilon]\colon (s' = 2);$$

This makes it straightforward to adapt existing DTMC or MDP benchmarks [164] to their interval variants, as done for example in [193].

PRISM provides *robust* verification, quantifying over all possible transition probabilities contained within the models' uncertainty sets. Property specification extends the existing PRISM property language. For IDTMCs and IMDPs, the tool supports the temporal logic PCTL, extended with (expected) reward operators and (co-safe) LTL formulae. For example, formulas $P_{maxmin=?}[\text{F } goal]$ and $P_{maxmax=?}[\text{F } goal]$ ask for the worst- and best-case scenarios, respectively, for maximising the probability of reaching a *goal*-labelled state.

Like many probabilistic model checking implementations, the uncertain models are solved via dynamic programming, in this case, *robust value iteration* [187, 220], implemented in PRISM's Java-based explicit-state model checking engine. Optimal policies for IMDPs can be generated and exported or simulated. Access to IDTMC and IMDP model checking is also provided programmatically at an API level, and has been applied to various problems, including anytime model learning [210] and abstraction of dynamical systems [19].

10.2   Outlook

Tool support for uncertainty models can be extended in various directions, for example to provide model checking for some of the model classes identified in Fig. 6 featuring partial observability (uncertain POMDPs) or adversarial behaviour (uncertain SGs), as well as improving efficiency and scalability for the simpler model classes. It will also be beneficial to extend the range of uncertainty types beyond intervals, which also necessitates more significant modelling language extensions.

11   State Space Exploration

State space exploration engines form the foundation of numerous quantitative analysis tools, playing a pivotal role in their functionality. Explicit-state model checkers, such as STORM with its **sparse** engine and MCSTA, rely on exploration engines to exhaustively construct the complete state space of a model before applying probabilistic model checking algorithms. Additionally, statistical model checkers such as MODES leverage exploration engines to generate large amounts of traces for statistical analysis. Exploration engines have recently also been used for training and verifying reinforcement learning agents [97,99].

In an effort to better understand the performance characteristics of the exploration engines utilised in different tools, we systematically benchmark and compare them. For this purpose, we consider the time and space needed for building an explicit representation of the complete state space of a model. Additionally, we compare the engines based on qualitative criteria such as the types of models they can handle and the interfaces they provide.

11.1   Tool Support

The tools participating in this category are the MODEST TOOLSET, MOMBA, and STORM. Both MOMBA and STORM participate with multiple engines, adding further diversity to the evaluation. Since all three tools support JANI, we employ it as a foundation for comparing and contrasting their capabilities.

The MODEST TOOLSET includes a state space exploration engine written in C# that is used by several of its tools, including MCSTA and MODES. It supports all types of models specified by JANI, including all JANI extensions. In that regard, it stands out as the most versatile among the engines we consider. For PTA, the engine supports the digital clock semantics [165], explicit valuations, clock regions [120], as well as clock zones [70]. In addition, it supports a symbolic treatment of continuous variables for hybrid models. In contrast to both STORM and MOMBA, which both provide public interfaces to their engines, the MODEST TOOLSET's engine is intended for internal use only and does not provide a public interface. The MODEST TOOLSET includes a separate MOPY transpilation tool to convert models to Python code implementing a first-state-next-state interface which can be used to explore the model's state space. In our experiments below, we access the MODEST TOOLSET's state space exploration engine via MCSTA.

MOMBA includes as a key feature a state space exploration engine designed to make exploration readily accessible via a comprehensive Python API. To achieve good performance, the engine is written in Rust. While MOMBA itself supports all of JANI, its state space exploration engine is more limited: It supports all discrete-time model types and flavours of timed automata specified by JANI except stochastic timed automata. The supported JANI extensions are `arrays`, `derived-operators`, `named-expressions`, and `trigonometric-functions`. In particular, the `functions` extension is not supported yet. For timed automata, it supports explicit valuations as well as clock zones. The Python API also provides functionality that goes beyond mere exploration: for instance, arbitrary JANI expressions can be evaluated in a given state and, for timed automata, clock zones can be manipulated. In addition to its traditional state space exploration engine, MOMBA also participates with an experimental new engine supporting a parallelized exploration mode harnessing the potential of multi-core systems. This experimental engine does not currently support timed automata and is not yet exposed via the Python API.

STORM participates with its `sparse` and `dd-to-sparse` engines. While STORM's `sparse` engine, like the engines of the MODEST TOOLSET and MOMBA, adopts a conventional explicit-state approach, the `dd-to-sparse` engine is based on first constructing a symbolic representation using BDDs of the state space and subsequently translating this to a traditional explicit representation. STORM supports all discrete- and continuous-time model types specified by JANI, except timed and hybrid automata. The supported JANI extensions are `arrays`, `derived-operators`, `functions`, and `state-exit-rewards`. STORM provides both a C++ and a Python interface, the latter as part of STORMPY, to its state space exploration engine. While the C++ API is fully featured, the Python API only supports the exploration of the entire state space of JANI models (but not the simulation of individual traces) while it has no such limitation for PRISM models. In contrast to Modest Toolset and MOMBA, STORM offers support for arbitrary-precision arithmetic using rational numbers implemented in the GMP library. This enables precise calculations and analysis, particularly when dealing with models that require high precision.

11.2   Performance Comparison

In our experimental evaluation, we utilise the QVBS as the foundation for benchmarking the tools. To ensure a meaningful comparison, we focus exclusively on discrete-time models, as these are supported by all the participating tools. Out of our initial 229 QVBS benchmarks, 25 resulted in timeouts after 30 min or were unsupported by all tools. Hence, the following analysis focuses on the remaining 204 benchmarks. For each benchmark, we measure the time required by each state space exploration engine to construct the entire state space. Additionally, we track the number of states counted by the engines and assess the memory consumption associated with each state where applicable. All benchmarks ran on a computer equipped with a 16-core AMD EPYC-Milan processor running at 3.4 GHz and 128 GB of RAM.

Table 9. Number of benchmarks per outcome and state space exploration engine

Engine	solved	unsupported	timeout	error
MODEST TOOLSET	194	9	1	0
MOMBA (v1)	159	45	0	0
MOMBA (v2,seq)	159	45	0	0
MOMBA (v2,par)	154	45	5	0
STORM (dd-to-sparse)	195	3	2	4
STORM (sparse)	202	0	2	0

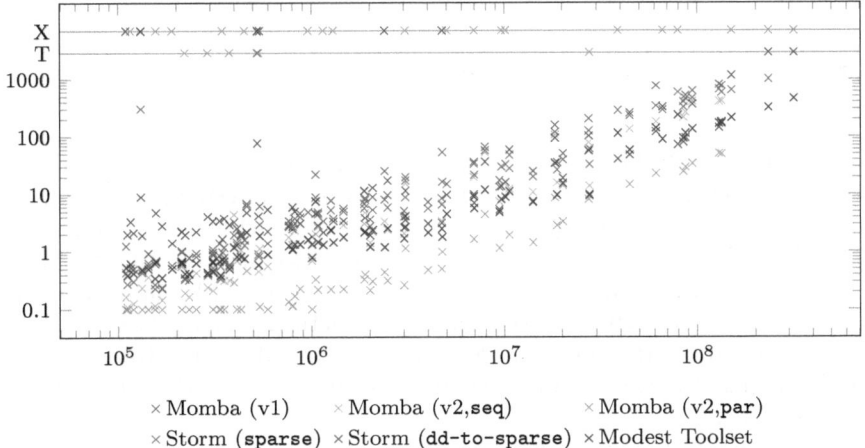

× Momba (v1)   × Momba (v2,seq)   × Momba (v2,par)
× Storm (sparse) × Storm (dd-to-sparse) × Modest Toolset

Fig. 7. Runtimes in seconds in relation to the total number of states

Table 9 shows the number of benchmarks per tool and our experiments' qualitative outcomes: we display the number of benchmarks that were successfully *solved, unsupported,* lead to a *timeout,* or caused an *error.* The 9 benchmarks not supported by the MODEST TOOLSET's engine use a complex specification for the initial states. The 45 benchmarks not supported by MOMBA use the functions JANI extension and are a superset of the 9 benchmarks not supported by the MODEST TOOLSET. The 3 benchmarks not supported by STORM's dd-to-sparse engine use assignment indices while for 4 benchmarks the same engine returned an error due to the BDD implementation running out of memory. The timeouts are all for different benchmarks. While the number of states reported by STORM and MOMBA is the same for all benchmarks and engines, the MODEST TOOLSET sometimes reports fewer states which presumably is due to some state space-reducing optimizations.

Runtimes. Figure 7 depicts the running time for each benchmark (on the vertical axis) in relation to the total number of states of the respective benchmark (on

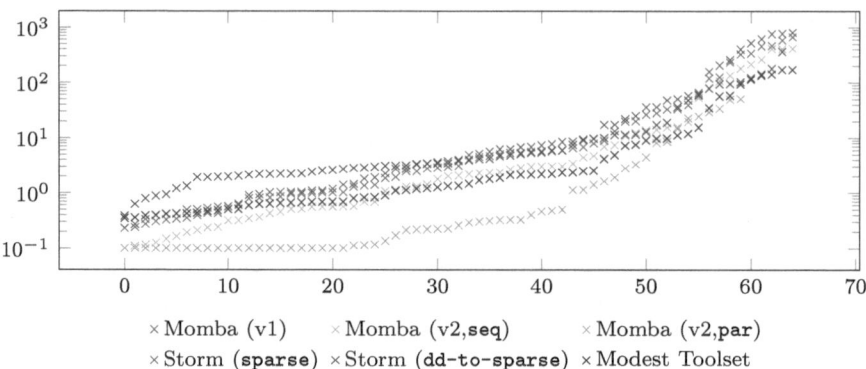

Fig. 8. Runtimes (s) vs. the number of benchmarks each solved in that time

the horizontal axis). The marks at the top indicate timeouts (T), and unsupported benchmarks as well as benchmarks returning an error (X). The quantile plot in Fig. 8 presents the cumulative number of benchmarks solved within a certain time. For presentation purposes, we chose to clamp the running times at $0.1\,s$ and restrict the plots to benchmarks with more than 10^5 states. For smaller benchmarks, the differences in runtimes are practically insignificant. Additionally, Fig. 8 is restricted to benchmarks supported by all engines to prevent skewing the plot (as otherwise an unsupported benchmark and a timeout would have the same effect).

From these results, it is evident that the approach taken by the `dd-to-sparse` engine of STORM only pays off for larger models; even then, it is rarely faster than the conventional explicit engine of the MODEST TOOLSET. Among those engines exclusively using a single core, the MODEST TOOLSET engine is almost always the fastest, although it has a larger startup overhead. This does not come as a surprise because, for efficiency, it is based on compiling JANI models to C# bytecode that is JIT-compiled. STORM's `dd-to-sparse` engine, like MOMBA's experimental parallel engine (v2,`par`), uses multiple cores since the underlying BDD implementation in SYLVAN [75] is parallelised. MOMBA's parallel engine is always faster than any other engine for benchmarks of a significant size. The average speed-up when compared to its sequential version is a factor of 9.1. In general, though, the runtimes of all engines are often quite similar.

Note that, as STORM and MCSTA are model checkers, they do a bit more work than MOMBA by creating a sparse matrix representation of the transitions and computing atomic propositions. We expect the performance impact of this to be minor—however we did not measure it.

Memory Consumption. Another interesting dimension when it comes to state space construction is the required memory. Efficiency is crucial here given the often huge state spaces due to the state space explosion problem. For the traditional explicit state engines, the size of the state space is linear in the number

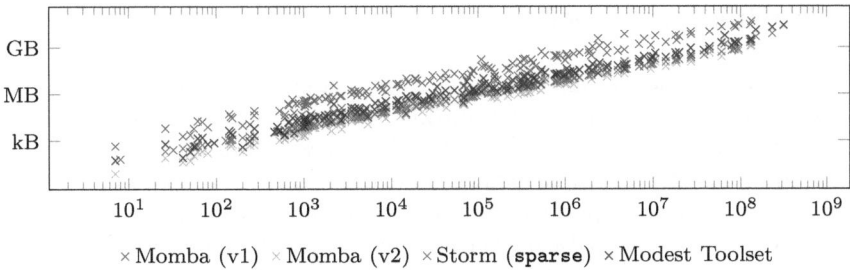

×Momba (v1) ×Momba (v2) ×Storm (`sparse`) ×Modest Toolset

Fig. 9. Size of the state space in relation to the number of states

of states. Figure 9 shows the size of the state spaces in relation to the number of states, computed based on the number of states and the size of each state. Note that the sequential and parallel variant of MOMBA's experimental engine use the same representation. In contrast to the MODEST TOOLSET, STORM's `sparse` engine and MOMBA's experimental engine use a more space efficient bit-packing representation of states, thereby reducing the amount of required memory. MOMBA's original engine uses the worst representation and always requires at least 16 bytes per variable independent of its actual domain.

Summary. Our results show that all engines are roughly comparable with respect to the time it takes to construct the entire state space of a model. STORM's `dd-to-sparse` engine may only be advantageous in terms of runtime for some large models while incurring a high overhead for small models. Among single-core engines, the MODEST TOOLSET's engine is almost always the fastest, especially for large models, while being the most versatile at the same time. The experimental parallel engine of MOMBA demonstrates that parallel state space exploration can be highly beneficial for larger models. The original MOMBA engine requires significantly more memory than all others. The MODEST TOOLSET's engine, however, does not provide a public API. Thus, if integration into another tool is a concern, STORM and, in particular, MOMBA with its original engine have an advantage as they both provide a Python API in addition to APIs in C++ and Rust, respectively.

Limitations. One of the motivations of this category is the lack assessment for simulation of individual traces. Note that the performance characteristics displayed here may not carry over to simulation of individual traces as there is a difference between always computing all successor states, as required for exhaustive exploration, and selectively computing only individual successor states which is, for instance, explicitly supported by MOMBA. Additionally, an exhaustive exploration requires maintaining a (hash) set of all visited states.

Data Availability. An artifact allowing to reproduce the performance comparison is archived and available at DOI 10.5281/zenodo.10626177 [144].

12   Stochastic Games

Game theory provides an effective way to model strategic interactions between multiple agents (or players) collaborating or competing to achieve objectives. Games have long been of interest within formal verification, providing a natural way to model, for example, honest and malicious participants in a security protocol or a controller operating in an adversarial environment.

In the context of quantitative verification, *stochastic games* (SGs) are a natural model to reason about strategic interactions in the context of uncertainty, noise, or randomisation. Verification problems for SGs have been studied for over 20 years [57]; the first model checking tools for SGs appeared over 10 years ago [63], and there has been growing interest in the topic recently. In essence, SGs (visualised on the right) generalise MDPs by permitting multiple players to have distinct strategies about how to resolve nondeterministic choices in the model. The simplest model, a turn-based SG (TSG), simply partitions the state space of an MDP, with the choices in each state being under the control of exactly one player. A concurrent SG (CSG) provides a more realistic model of concurrent decision-making: in each state, players resolve their choices independently.

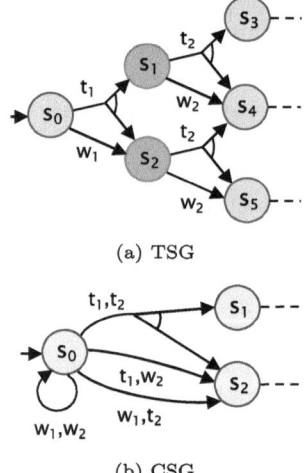

Verification of SGs also takes a variety of flavours. The simplest option is a *zero-sum* setting, where one player (or a coalition of players) aims to maximise some objective, such as the probability of reaching a set of target states or satisfying a temporal logic formula, and the other player (or players) have the opposite objective, i.e. to minimise it. For SGs, the logic rPATL [63] is widely used, which generalises the well-known game logic ATL [5] to a variety of quantitative objectives. Beyond zero-sum properties, temporal logics and model checking algorithms have been extended [159] to support *equilibria*, which are joint strategies where each player optimises their own distinct objective in such a way that it is not beneficial for any player to unilaterally change strategy.

12.1   Algorithms and Tool Support

Despite verification problems for SGs typically having a higher complexity than their MDP counterparts, core properties of TSGs can be effectively analysed with similar methods such as value iteration [66] or interval iteration and its variants [13,80]. Methods to solve CSGs tend to be more expensive: again they are usually based on value iteration, but require the solution of a linear programming or equilibrium synthesis problem [159] for every state at each iteration.

Verification tools for SGs under active development are PRISM-GAMES and its extensions, TEMPEST, PET, and EPMC. We provide a brief empirical comparison of the first four below. These tools share a common input format for SGs, namely the PRISM-GAMES modelling language. This extends the widely used PRISM modelling language: In the case of TSGs, it is a rather simple extension of the case for MDPs, defining a set of players and the states they own; CSGs use a different model of parallel composition and additional language features.

EPMC also supports the analysis of stochastic parity games and verification of epistemic properties on probabilistic multi-agent systems in addition to its standard probabilistic model checking functionality.

PET has recently been extended to support reachability objectives for TSGs. It uses PRISM-GAMES to parse and explore games, and employs the interval iteration approach of [80] to solve them. Implementing partial exploration based on [80, Sect. 5] in combination with the approach of [152] for more complex objectives such as total reward or mean payoff is planned.

PRISM-GAMES mainly focuses on TSGs and CSGs, but it also supports turn-based probabilistic timed games. The tool supports a wide range of zero-sum properties (probabilistic reachability, expected rewards, co-safe linear temporal logic and multi-objective specifications) as well as (social welfare) Nash equilibria. Recent extensions add support for correlated equilibria and social fairness [160]. The implementation is primarily based on variants of value iteration, implemented in Java with explicit state data structures, but also includes symbolic (MTBDD-based) model checking of TSGs [161].

TEMPEST extends STORM to TSGs with a focus on synthesizing most-permissive strategies. The tool supports zero-sum properties, namely probabilistic reachability and mean-payoff properties. The model checking procedures are based on variants of value iteration using explicit representations of the state space.

[154] and [13] present an extension of PRISM-GAMES which adds various methods for solving TSGs: interval iteration (II) [80] and optimistic value iteration (OVI) [13], as well as topological variants of each; the "widest path" (WP) variant of II [190]; and solution methods based on strategy iteration and quadratic programming. The latter are omitted from our comparison since they are fundamentally different from the variants of value iteration employed by the other tools [154, Sect. 5.5.3]; we refer to [154, Sect. 5] for a practical comparison of these solution methods.

Also relevant are GIST [58] and GAVS+ [65], which implement TSG verification, but are no longer developed or maintained, and UPPAAL STRATEGO [72], which supports stochastic priced timed games via multiple other UPPAAL branches.

Table 10. Performance comparison results of tools for stochastic games

Benchmark			Value iteration (s)			ε-exact (s)			
Model + property [parameters]	Param. values	# states	PRISM -GAMES (expl.)	PRISM -GAMES (symb.)	TEMP -EST	PET	P-G+ (II)	P-G+ (OVI)	P-G+ (WP)
avoid + find [X_MAX, Y_MAX]	10, 10	106,524	16.9	15.4	**1.4**	**5.0**	17.2	22.4	16.7
	15, 15	480,464	125.9	62.6	**4.7**	**15.7**	126.9	137.2	126
	20, 20	1,436,404	T/O	240.8	**12.9**	**57.5**	T/O	T/O	T/O
hallway_human + save [X_MAX, Y_MAX]	5, 5	25,000	2.5	1.8	**0.9**	2.9	**2.4**	**2.4**	**2.4**
	10, 10	400,000	10.5	**2.0**	12.9	**9.5**	11.3	11.2	11.3
	15, 15	2,025,000	50.1	**4.0**	101.3	**39.6**	57.0	55.4	56.6
investors + greater [N, vmax]	2, 20	568,790	21.8	7.4	**4.9**	**16.7**	33.2	42.3	54.6
	2, 40	2,041,690	98.8	26.0	**19.8**	**69.0**	144.8	183.2	314.6
	3, 20	4,058,751	167.7	**19.2**	39.7	**152.6**	241.4	321.3	484.8
	3, 40	14,569,251	M/O	**62.8**	171.2	T/O	T/O	T/O	T/O
safe_nav + reach [N, feat]	8, D	2,592,845	544.2	**16.2**	518.5	519.4	498.4	508.7	**485.7**
	8, A	17,052,941	T/O	**110.8**	T/O	T/O	T/O	T/O	T/O
BigMec + BigMec [N]	10,000	20,003	46.9	9.3	**2.5**	17.6	T/O	49.5	73.5
	25,000	50,003	290.4	45.8	**12.7**	82.9	T/O	294.2	472.4
ManyMec + ManyMec [N]	10,000	30,002	160.7	263.3	**16.7**	104.3	T/O	T/O	460.4
	25,000	75,002	T/O	T/O	**98.6**	T/O	T/O	T/O	T/O

12.2 Performance Comparison

We give a brief performance comparison of the various tools and techniques, focusing on the problem class supported by all tools: zero-sum probabilistic reachability for TSGs. Table 10 shows total runtimes (game construction and solution) on an indicative set of benchmarks from the PRISM Benchmark Suite [164] and [154]. Experiments ran on an AMD Ryzen 5 3600 system, pinned to a single core and restricted to 8 GB of RAM, running inside Docker, using OpenJDK JRE-17 for all Java tools, and with a timeout (T/O) of 10 min. For each invocation, a fresh docker container is created.

For a fair comparison, we group them into two distinct categories based on the degree of accuracy provided: "value iteration" (i.e. no strict guarantees on the correctness of the result) and "ε-exact" (the result is guaranteed to be within $\pm \varepsilon = 10^{-6}$ of the true value), marking the fastest tool in each category in bold.

Value Iteration. Comparing explicit-state implementations, TEMPEST is faster than PRISM-GAMES on almost all instances (primarily, it appears, due to the former being implemented in C++, but the latter also uses slower but more extensive precomputations). PRISM-GAMES, in symbolic mode, outperforms TEMPEST on most larger models and scales to the biggest TSGs of all tools. Symbolic model building times (not shown) are also usually faster.

ε-exact. PET outperforms the approaches in the PRISM-GAMES extension of [13, 154] (denoted P-G+ in the table) on practically all models. This is interesting since the algorithmic approach in the former is the same as interval iteration (II) in the latter. Since these tools are implemented in the same language (Java) and use the same model construction (PRISM's model generator), the (significant) differences are solely a result of engineering. Times for the methods in the PRISM-GAMES extension are typically in the same order of magnitude, however there are models where one approach significantly outperforms all others.

Data Availability. All tools, models and scripts needed to replicate our results can be found at DOI 10.5281/zenodo.7831387 [180].

12.3 Outlook

Interest has grown in the formal verification of SGs in recent years and it has already been applied to a range of domains, from computer security to adaptive software architectures (as evidenced by the collection of PRISM-GAMES case studies at prismmodelchecker.org/games/casestudies.php). In addition to improving the efficiency and scalability of existing tools, one key challenge is to develop methods for **partially observable** variants of SG models. Another is to develop support for **richer specification languages**, for example incorporating strategies, equilibria or epistemic properties.

Table 11. Data availability for QComp 2023

Section	Category	DOI	Ref.
4	Long-Run Average Rewards	10.5281/zenodo.8219191	[100]
6	Multi-Objective Analysis	10.5281/zenodo.8063883	[195]
7	Parametric Markov Models	10.5281/zenodo.10646479	[137]
8	Partially-Observable MDPs	10.5281/zenodo.8215337	[31]
9	Rare Events	10.6084/m9.figshare.23818395	[47]
11	State Space Exploration	10.5281/zenodo.10626177	[144]
12	Stochastic Games	10.5281/zenodo.7831387	[180]

13   Conclusion

We have described the state of the art in tools and algorithms at the frontiers of quantitative verification in ten different categories, covering 19 different tools. In several categories, we reported on the first systematic performance comparison among the included tools. On many of the frontiers we described, tool support for advanced properties and models is now being consolidated, but a plethora of open questions and unimplemented ideas remain for future work. We hope that this report can serve as an inspiration for further work on quantitative verification tooling, and that several of QComp 2023's categories can evolve into regular, serious performance evaluations among competing tools in the near future. At the same time, it is clear that our coverage of the quantitative verification frontiers is not complete. As one example, we mention the area of parametric models based on timed automata (in which parameters are traditionally more structural in nature than the ones in the parametric Markov models of Sect. 7) where tools are maturing [6] and benchmark sets with support for JANI are being collected [7], laying the foundations for future performance evaluations.

For the next edition of QComp, which at the latest will take place in time for the next edition of the TOOLympics, we intend to keep the multiple-category setup. We plan to both add new categories, e.g. on parametric timed automata as mentioned above or on entirely new problems that surface in the coming years, and also perform more extensive performance evaluations in those categories where tools will have matured sufficiently and a good benchmark set will have become available. As such, we expect a mix of "friendly" categories that stimulate tool development and standardisation as well as more "competitive" evaluations where performance really counts. Practically, we may need to split off the reports of the larger categories—those where many tools are evaluated on comprehensive benchmark sets to obtain representative performance comparisons—from the main competition report into publications of their own. In parallel to the transformation of QComp that started with this edition, the comparison of established tools on basic problems as in QComp 2019 and 2020 is likely to transition into a continuous evaluation—rather than periodic competitions—hosted on the qcomp.org website. We look forward to a continuing journey into the undiscovered country beyond today's frontiers of quantitative verification in the next editions of the QComp friendly competition!

Data Availability. In each category that performed a performance comparison, we provide an artifact that archives the models, tools, scripts, and other data that is necessary to reproduce the respective experiments. The benchmark set of parametric Markov models introduced in Sect. 7 is also publicly archived. We link to the DOIs of the respective datasets at the end of each of Sects. 4, 6, 7, 8, 9, 11, and 12, and list all of them in Table 11.

References

1. Abate, A., Andriushchenko, R., Ceska, M., Kwiatkowska, M.: Adaptive formal approximations of Markov chains. Perform. Eval. **148**, 102207 (2021). https://doi.org/10.1016/j.peva.2021.102207

2. Agarwal, C., Guha, S., Kretínský, J., Muruganandham, P.: PAC statistical model checking of mean payoff in discrete- and continuous-time MDP. In: Shoham, S., Vizel, Y. (eds.) CAV 2022. LNCS, vol. 13372, pp. 3–25. Springer, Cham (2022). https://doi.org/10.1007/978-3-031-13188-2_1

3. Agha, G., Palmskog, K.: A survey of statistical model checking. ACM Trans. Model. Comput. Simul. **28**(1), 6:1–6:39 (2018). https://doi.org/10.1145/3158668

4. Alur, R., Dill, D.L.: A theory of timed automata. Theor. Comput. Sci. **126**(2), 183–235 (1994). https://doi.org/10.1016/0304-3975(94)90010-8

5. Alur, R., Henzinger, T.A., Kupferman, O.: Alternating-time temporal logic. J. ACM **49**(5), 672–713 (2002). https://doi.org/10.1145/585265.585270

6. André, É.: IMITATOR 3: synthesis of timing parameters beyond decidability. In: Silva, A., Leino, K.R.M. (eds.) CAV 2021. LNCS, vol. 12759, pp. 552–565. Springer, Cham (2021). https://doi.org/10.1007/978-3-030-81685-8_26

7. André, É., Marinho, D., van de Pol, J.: A benchmarks library for extended parametric timed automata. In: Loulergue, F., Wotawa, F. (eds.) TAP 2021. LNCS, vol. 12740, pp. 39–50. Springer, Cham (2021). https://doi.org/10.1007/978-3-030-79379-1_3

8. Andriushchenko, R., Ceska, M., Junges, S., Katoen, J.P.: Inductive synthesis of finite-state controllers for POMDPs. In: Cussens, J., Zhang, K. (eds.) 38th Conference on Uncertainty in Artificial Intelligence (UAI). Proceedings of Machine Learning Research, vol. 180, pp. 85–95. PMLR (2022)

9. Andriushchenko, R., Češka, M., Junges, S., Katoen, J.-P., Stupinský, Š: PAYNT: a tool for inductive synthesis of probabilistic programs. In: Silva, A., Leino, K.R.M. (eds.) CAV 2021. LNCS, vol. 12759, pp. 856–869. Springer, Cham (2021). https://doi.org/10.1007/978-3-030-81685-8_40

10. Arming, S., Bartocci, E., Chatterjee, K., Katoen, J.-P., Sokolova, A.: Parameter-independent strategies for pMDPs via POMDPs. In: McIver, A., Horvath, A. (eds.) QEST 2018. LNCS, vol. 11024, pp. 53–70. Springer, Cham (2018). https://doi.org/10.1007/978-3-319-99154-2_4

11. Ashok, P., Brázdil, T., Křetínský, J., Slámečka, O.: Monte Carlo tree search for verifying reachability in Markov decision processes. In: Margaria, T., Steffen, B. (eds.) ISoLA 2018. LNCS, vol. 11245, pp. 322–335. Springer, Cham (2018). https://doi.org/10.1007/978-3-030-03421-4_21

12. Ashok, P., Chatterjee, K., Daca, P., Křetínský, J., Meggendorfer, T.: Value iteration for long-run average reward in Markov decision processes. In: Majumdar, R., Kunčak, V. (eds.) CAV 2017. LNCS, vol. 10426, pp. 201–221. Springer, Cham (2017). https://doi.org/10.1007/978-3-319-63387-9_10

13. Azeem, M., Evangelidis, A., Kretínský, J., Slivinskiy, A., Weininger, M.: Optimistic and topological value iteration for simple stochastic games. In: Bouajjani, A., Holík, L., Wu, Z. (eds.) ATVA 2022. LNCS, vol. 13505, pp. 285–302. Springer, Cham (2022). https://doi.org/10.1007/978-3-031-19992-9_18

14. Aziz, A., Sanwal, K., Singhal, V., Brayton, R.K.: Model-checking continous-time Markov chains. ACM Trans. Comput. Log. **1**(1), 162–170 (2000). https://doi.org/10.1145/343369.343402

15. Babiak, T., et al.: The Hanoi omega-automata format. In: Kroening, D., Păsăreanu, C.S. (eds.) CAV 2015. LNCS, vol. 9206, pp. 479–486. Springer, Cham (2015). https://doi.org/10.1007/978-3-319-21690-4_31

16. Backenköhler, M., Bortolussi, L., Großmann, G., Wolf, V.: Abstraction-guided truncations for stationary distributions of Markov population models. In: Abate, A., Marin, A. (eds.) QEST 2021. LNCS, vol. 12846, pp. 351–371. Springer, Cham (2021). https://doi.org/10.1007/978-3-030-85172-9_19

17. Badings, T., Simão, T.D., Suilen, M., Jansen, N.: Decision-making under uncertainty: beyond probabilities. Int. J. Softw. Tools Technol. Transf. (2023). https://doi.org/10.1007/s10009-023-00704-3

18. Badings, T.S., Cubuktepe, M., Jansen, N., Junges, S., Katoen, J.P., Topcu, U.: Scenario-based verification of uncertain parametric MDPs. Int. J. Softw. Tools Technol. Transf. **24**(5), 803–819 (2022). https://doi.org/10.1007/s10009-022-00673-z

19. Badings, T.S., et al.: Robust control for dynamical systems with non-Gaussian noise via formal abstractions. J. Artif. Intell. Res. **76**, 341–391 (2023). https://doi.org/10.1613/jair.1.14253

20. Baier, C., de Alfaro, L., Forejt, V., Kwiatkowska, M.: Model checking probabilistic systems. In: Handbook of Model Checking, pp. 963–999. Springer, Cham (2018). https://doi.org/10.1007/978-3-319-10575-8_28

21. Baier, C., Haverkort, B.R., Hermanns, H., Katoen, J.P.: Model-checking algorithms for continuous-time Markov chains. IEEE Trans. Software Eng. **29**(6), 524–541 (2003). https://doi.org/10.1109/TSE.2003.1205180

22. Baier, C., Hensel, C., Hutschenreiter, L., Junges, S., Katoen, J.P., Klein, J.: Parametric Markov chains: PCTL complexity and fraction-free Gaussian elimination. Inf. Comput. **272**, 104504 (2020). https://doi.org/10.1016/j.ic.2019.104504

23. Bals, S., Evangelidis, A., Grover, K., Kretínský, J., Waibel, J.: MULTIGAIN 2.0: MDP controller synthesis for multiple mean-payoff, LTL and steady-state constraints. CoRR **abs/2305.16752** (2023). https://doi.org/10.48550/arXiv.2305.16752

24. Barbot, B., Haddad, S., Picaronny, C.: Coupling and importance sampling for statistical model checking. In: Flanagan, C., König, B. (eds.) TACAS 2012. LNCS, vol. 7214, pp. 331–346. Springer, Heidelberg (2012). https://doi.org/10.1007/978-3-642-28756-5_23

25. Bartocci, E., et al.: TOOLympics 2019: an overview of competitions in formal methods. In: Beyer, D., Huisman, M., Kordon, F., Steffen, B. (eds.) TACAS 2019. LNCS, vol. 11429, pp. 3–24. Springer, Cham (2019). https://doi.org/10.1007/978-3-030-17502-3_1

26. Bartocci, E., Grosu, R., Katsaros, P., Ramakrishnan, C.R., Smolka, S.A.: Model repair for probabilistic systems. In: Abdulla, P.A., Leino, K.R.M. (eds.) TACAS 2011. LNCS, vol. 6605, pp. 326–340. Springer, Heidelberg (2011). https://doi.org/10.1007/978-3-642-19835-9_30

27. Basset, N., Kwiatkowska, M., Topcu, U., Wiltsche, C.: Strategy synthesis for stochastic games with multiple long-run objectives. In: Baier, C., Tinelli, C. (eds.) TACAS 2015. LNCS, vol. 9035, pp. 256–271. Springer, Heidelberg (2015). https://doi.org/10.1007/978-3-662-46681-0_22

28. Batz, K., Junges, S., Kaminski, B.L., Katoen, J.-P., Matheja, C., Schröer, P.: PrIC3: property directed reachability for MDPs. In: Lahiri, S.K., Wang, C. (eds.) CAV 2020. LNCS, vol. 12225, pp. 512–538. Springer, Cham (2020). https://doi.org/10.1007/978-3-030-53291-8_27

29. Bellman, R.: A Markovian decision process. J. Math. Mech. **6**(5), 679–684 (1957)
30. Bohnenkamp, H.C., D'Argenio, P.R., Hermanns, H., Katoen, J.P.: MoDeST: a compositional modeling formalism for hard and softly timed systems. IEEE Trans. Software Eng. **32**(10), 812–830 (2006). https://doi.org/10.1109/TSE.2006.104
31. Bork, A.: Replication package QComp 2023 – POMDP analysis (2023). https://doi.org/10.5281/zenodo.8215337
32. Bork, A., Junges, S., Katoen, J.-P., Quatmann, T.: Verification of indefinite-horizon POMDPs. In: Hung, D.V., Sokolsky, O. (eds.) ATVA 2020. LNCS, vol. 12302, pp. 288–304. Springer, Cham (2020). https://doi.org/10.1007/978-3-030-59152-6_16
33. Bork, A., Katoen, J.P., Quatmann, T.: Under-approximating expected total rewards in POMDPs. In: Fisman, D., Rosu, G. (eds.) TACAS 2022. LNCS, vol. 13244, pp. 22–40. Springer, Cham (2022). https://doi.org/10.1007/978-3-030-99527-0_2
34. Bortolussi, L., Silvetti, S.: Bayesian statistical parameter synthesis for linear temporal properties of stochastic models. In: Beyer, D., Huisman, M. (eds.) TACAS 2018. LNCS, vol. 10806, pp. 396–413. Springer, Cham (2018). https://doi.org/10.1007/978-3-319-89963-3_23
35. Brázdil, T., Brozek, V., Chatterjee, K., Forejt, V., Kucera, A.: Two views on multiple mean-payoff objectives in Markov decision processes. Log. Methods Comput. Sci. **10**(1) (2014). https://doi.org/10.2168/LMCS-10(1:13)2014
36. Brázdil, T., et al.: Verification of Markov decision processes using learning algorithms. In: Cassez, F., Raskin, J.-F. (eds.) ATVA 2014. LNCS, vol. 8837, pp. 98–114. Springer, Cham (2014). https://doi.org/10.1007/978-3-319-11936-6_8
37. Brázdil, T., Chatterjee, K., Forejt, V., Kučera, A.: MULTIGAIN: a controller synthesis tool for MDPs with multiple mean-payoff objectives. In: Baier, C., Tinelli, C. (eds.) TACAS 2015. LNCS, vol. 9035, pp. 181–187. Springer, Heidelberg (2015). https://doi.org/10.1007/978-3-662-46681-0_12
38. Brim, L., Češka, M., Dražan, S., Šafránek, D.: Exploring parameter space of stochastic biochemical systems using quantitative model checking. In: Sharygina, N., Veith, H. (eds.) CAV 2013. LNCS, vol. 8044, pp. 107–123. Springer, Heidelberg (2013). https://doi.org/10.1007/978-3-642-39799-8_7
39. Bry, A., Roy, N.: Rapidly-exploring random belief trees for motion planning under uncertainty. In: 2011 IEEE International Conference on Robotics and Automation (ICRA), pp. 723–730. IEEE (2011). https://doi.org/10.1109/ICRA.2011.5980508
40. Budde, C.E., D'Argenio, P.R., Hartmanns, A.: Automated compositional importance splitting. Sci. Comput. Program. **174**, 90–108 (2019). https://doi.org/10.1016/j.scico.2019.01.006
41. Budde, C.E., D'Argenio, P.R., Hartmanns, A., Sedwards, S.: An efficient statistical model checker for nondeterminism and rare events. Int. J. Softw. Tools Technol. Transf. **22**(6), 759–780 (2020). https://doi.org/10.1007/s10009-020-00563-2
42. Budde, C.E., D'Argenio, P.R., Monti, R.E.: Compositional construction of importance functions in fully automated importance splitting. In: Puliafito, A., Trivedi, K.S., Tuffin, B., Scarpa, M., Machida, F., Alonso, J. (eds.) 10th EAI International Conference on Performance Evaluation Methodologies and Tools (VALUETOOLS). ACM (2016). https://doi.org/10.4108/eai.25-10-2016.2266501
43. Budde, C.E., D'Argenio, P.R., Monti, R.E., Stoelinga, M.: Analysis of non-Markovian repairable fault trees through rare event simulation. Int. J. Softw. Tools Technol. Transf. **24**(5), 821–841 (2022). https://doi.org/10.1007/s10009-022-00675-x

44. Budde, C.E., Dehnert, C., Hahn, E.M., Hartmanns, A., Junges, S., Turrini, A.: JANI: quantitative model and tool interaction. In: Legay, A., Margaria, T. (eds.) TACAS 2017. LNCS, vol. 10206, pp. 151–168. Springer, Heidelberg (2017). https://doi.org/10.1007/978-3-662-54580-5_9
45. Budde, C.E., Hartmanns, A.: Replicating *Restart* with prolonged retrials: an experimental report. In: TACAS 2021. LNCS, vol. 12652, pp. 373–380. Springer, Cham (2021). https://doi.org/10.1007/978-3-030-72013-1_21
46. Budde, C.E., et al.: On correctness, precision, and performance in quantitative verification. In: Margaria, T., Steffen, B. (eds.) ISoLA 2020. LNCS, vol. 12479, pp. 216–241. Springer, Cham (2021). https://doi.org/10.1007/978-3-030-83723-5_15
47. Budde, C.E., et al.: QComp 2023: formal tools for rare events (experimental reproduction package). Figshare (2023). https://doi.org/10.6084/m9.figshare.23818395
48. Budde, C.E., Ruijters, E., Stoelinga, M.: The dynamic fault tree rare event simulator. In: Gribaudo, M., Jansen, D.N., Remke, A. (eds.) QEST 2020. LNCS, vol. 12289, pp. 233–238. Springer, Cham (2020). https://doi.org/10.1007/978-3-030-59854-9_17
49. Buecherl, L., et al.: A collection of biological models for the development of infinite-state stochastic model checking tools. In: 15th International Workshop on Bio-Design Automation (IWBDA), pp. 44–47 (2023)
50. Burns, B., Brock, O.: Sampling-based motion planning with sensing uncertainty. In: 2007 IEEE International Conference on Robotics and Automation (ICRA), pp. 3313–3318. IEEE (2007). https://doi.org/10.1109/ROBOT.2007.363984
51. Butkova, Y., Hartmanns, A., Hermanns, H.: A modest approach to Markov automata. ACM Trans. Model. Comput. Simul. **31**(3), 14:1–14:34 (2021). https://doi.org/10.1145/3449355
52. Butkova, Y., Wimmer, R., Hermanns, H.: Long-run rewards for Markov automata. In: Legay, A., Margaria, T. (eds.) TACAS 2017. LNCS, vol. 10206, pp. 188–203. Springer, Heidelberg (2017). https://doi.org/10.1007/978-3-662-54580-5_11
53. Cardelli, L., Kwiatkowska, M., Laurenti, L.: A stochastic hybrid approximation for chemical kinetics based on the linear noise approximation. In: Bartocci, E., Lio, P., Paoletti, N. (eds.) CMSB 2016. LNCS, vol. 9859, pp. 147–167. Springer, Cham (2016). https://doi.org/10.1007/978-3-319-45177-0_10
54. Češka, M., Chau, C., Křetínský, J.: SeQuaiA: a scalable tool for semi-quantitative analysis of chemical reaction networks. In: Lahiri, S.K., Wang, C. (eds.) CAV 2020. LNCS, vol. 12224, pp. 653–666. Springer, Cham (2020). https://doi.org/10.1007/978-3-030-53288-8_32
55. Češka, M., Křetínský, J.: Semi-quantitative abstraction and analysis of chemical reaction networks. In: Dillig, I., Tasiran, S. (eds.) CAV 2019. LNCS, vol. 11561, pp. 475–496. Springer, Cham (2019). https://doi.org/10.1007/978-3-030-25540-4_28
56. Chatterjee, K., Gaiser, A., Křetínský, J.: Automata with generalized Rabin pairs for probabilistic model checking and LTL synthesis. In: Sharygina, N., Veith, H. (eds.) CAV 2013. LNCS, vol. 8044, pp. 559–575. Springer, Heidelberg (2013). https://doi.org/10.1007/978-3-642-39799-8_37
57. Chatterjee, K., Henzinger, T.A.: A survey of stochastic ω-regular games. J. Comput. Syst. Sci. **78**(2), 394–413 (2012). https://doi.org/10.1016/j.jcss.2011.05.002
58. Chatterjee, K., Henzinger, T.A., Jobstmann, B., Radhakrishna, A.: GIST: a solver for probabilistic games. In: Touili, T., Cook, B., Jackson, P. (eds.) CAV 2010. LNCS, vol. 6174, pp. 665–669. Springer, Heidelberg (2010). https://doi.org/10.1007/978-3-642-14295-6_57

59. Chatterjee, K., Katoen, J.P., Mohr, S., Weininger, M., Winkler, T.: Stochastic games with lexicographic objectives. Formal Methods Syst. Des. (2023). https://doi.org/10.1007/s10703-023-00411-4
60. Chatterjee, K., Katoen, J.-P., Weininger, M., Winkler, T.: Stochastic games with lexicographic reachability-safety objectives. In: Lahiri, S.K., Wang, C. (eds.) CAV 2020. LNCS, vol. 12225, pp. 398–420. Springer, Cham (2020). https://doi.org/10.1007/978-3-030-53291-8_21
61. Chatterjee, K., Kretínská, Z., Kretínský, J.: Unifying two views on multiple mean-payoff objectives in Markov decision processes. Log. Methods Comput. Sci. **13**(2) (2017). https://doi.org/10.23638/LMCS-13(2:15)2017
62. Chatterjee, K., Majumdar, R., Henzinger, T.A.: Markov decision processes with multiple objectives. In: Durand, B., Thomas, W. (eds.) STACS 2006. LNCS, vol. 3884, pp. 325–336. Springer, Heidelberg (2006). https://doi.org/10.1007/11672142_26
63. Chen, T., Forejt, V., Kwiatkowska, M., Parker, D., Simaitis, A.: Automatic verification of competitive stochastic systems. In: Flanagan, C., König, B. (eds.) TACAS 2012. LNCS, vol. 7214, pp. 315–330. Springer, Heidelberg (2012). https://doi.org/10.1007/978-3-642-28756-5_22
64. Chen, T., Hahn, E.M., Han, T., Kwiatkowska, M.Z., Qu, H., Zhang, L.: Model repair for Markov decision processes. In: Seventh International Symposium on Theoretical Aspects of Software Engineering (TASE), pp. 85–92. IEEE Computer Society (2013). https://doi.org/10.1109/TASE.2013.20
65. Cheng, C.-H., Knoll, A., Luttenberger, M., Buckl, C.: GAVS+: an open platform for the research of algorithmic game solving. In: Abdulla, P.A., Leino, K.R.M. (eds.) TACAS 2011. LNCS, vol. 6605, pp. 258–261. Springer, Heidelberg (2011). https://doi.org/10.1007/978-3-642-19835-9_22
66. Condon, A.: On algorithms for simple stochastic games. In: Cai, J.Y. (ed.) Advances in Computational Complexity Theory, Proceedings of a DIMACS Workshop. DIMACS Series in Discrete Mathematics and Theoretical Computer Science, vol. 13, pp. 51–71. DIMACS/AMS (1990). https://doi.org/10.1090/dimacs/013/04
67. Cubuktepe, M., Jansen, N., Junges, S., Katoen, J.P., Topcu, U.: Convex optimization for parameter synthesis in MDPs. IEEE Trans. Autom. Control **67**(12), 6333–6348 (2022). https://doi.org/10.1109/TAC.2021.3133265
68. Cubuktepe, M., Jansen, N., Junges, S., Marandi, A., Suilen, M., Topcu, U.: Robust finite-state controllers for uncertain POMDPs. In: 35th AAAI Conference on Artificial Intelligence (AAAI), pp. 11792–11800. AAAI Press (2021). https://doi.org/10.1609/aaai.v35i13.17401
69. Daigle, Bernie J., J., Roh, M.K., Gillespie, D.T., Petzold, L.R.: Automated estimation of rare event probabilities in biochemical systems. J. Chem. Phys. **134**(4) (2011). https://doi.org/10.1063/1.3522769
70. D'Argenio, P.R., Hartmanns, A., Legay, A., Sedwards, S.: Statistical approximation of optimal schedulers for probabilistic timed automata. In: Ábrahám, E., Huisman, M. (eds.) IFM 2016. LNCS, vol. 9681, pp. 99–114. Springer, Cham (2016). https://doi.org/10.1007/978-3-319-33693-0_7
71. D'Argenio, P.R., Monti, R.E.: Input/Output stochastic automata with urgency: confluence and weak determinism. In: Fischer, B., Uustalu, T. (eds.) ICTAC 2018. LNCS, vol. 11187, pp. 132–152. Springer, Cham (2018). https://doi.org/10.1007/978-3-030-02508-3_8

72. David, A., Jensen, P.G., Larsen, K.G., Mikucionis, M., Taankvist, J.H.: Uppaal Stratego. In: Baier, C., Tinelli, C. (eds.) TACAS 2015. LNCS, vol. 9035, pp. 206–211. Springer, Cham (2015). https://doi.org/10.1007/978-3-662-46681-0_16

73. Daws, C.: Symbolic and parametric model checking of discrete-time Markov chains. In: Liu, Z., Araki, K. (eds.) ICTAC 2004. LNCS, vol. 3407, pp. 280–294. Springer, Heidelberg (2005). https://doi.org/10.1007/978-3-540-31862-0_21

74. Delgrange, F., Katoen, J.-P., Quatmann, T., Randour, M.: Simple strategies in multi-objective MDPs. In: TACAS 2020. LNCS, vol. 12078, pp. 346–364. Springer, Cham (2020). https://doi.org/10.1007/978-3-030-45190-5_19

75. van Dijk, T., van de Pol, J.: Sylvan: multi-core framework for decision diagrams. Int. J. Softw. Tools Technol. Transf. **19**(6), 675–696 (2017). https://doi.org/10.1007/s10009-016-0433-2

76. Donovan, R.M., Sedgewick, A.J., Faeder, J.R., Zuckerman, D.M.: Efficient stochastic simulation of chemical kinetics networks using a weighted ensemble of trajectories. J. Chem. Phys. **139**(11) (2013). https://doi.org/10.1063/1.4821167

77. Duret-Lutz, A., et al.: From Spot 2.0 to Spot 2.10: What's new? In: Shoham, S., Vizel, Y. (eds.) CAV 2022. LNCS, vol. 13372, pp. 174–187. Springer, Cham (2022). https://doi.org/10.1007/978-3-031-13188-2_9

78. Egorov, M., Sunberg, Z.N., Balaban, E., Wheeler, T.A., Gupta, J.K., Kochenderfer, M.J.: POMDPs.jl: a framework for sequential decision making under uncertainty. J. Mach. Learn. Res. **18**, 26:1–26:5 (2017)

79. Eisentraut, C., Hermanns, H., Zhang, L.: On probabilistic automata in continuous time. In: 25th Annual IEEE Symposium on Logic in Computer Science (LICS), pp. 342–351. IEEE Computer Society (2010). https://doi.org/10.1109/LICS.2010.41

80. Eisentraut, J., Kelmendi, E., Kretínský, J., Weininger, M.: Value iteration for simple stochastic games: stopping criterion and learning algorithm. Inf. Comput. **285**(Part), 104886 (2022). https://doi.org/10.1016/j.ic.2022.104886

81. Esparza, J., Křetínský, J.: From LTL to deterministic automata: a safraless compositional approach. In: Biere, A., Bloem, R. (eds.) CAV 2014. LNCS, vol. 8559, pp. 192–208. Springer, Cham (2014). https://doi.org/10.1007/978-3-319-08867-9_13

82. Esparza, J., Kretínský, J., Sickert, S.: A unified translation of linear temporal logic to ω-automata. J. ACM **67**(6), 33:1–33:61 (2020). https://doi.org/10.1145/3417995

83. Etessami, K., Kwiatkowska, M.Z., Vardi, M.Y., Yannakakis, M.: Multi-objective model checking of Markov decision processes. Log. Methods Comput. Sci. **4**(4) (2008). https://doi.org/10.2168/LMCS-4(4:8)2008

84. Fedyukovich, G., Mover, S.: TACAS 23 artifact evaluation VM – Ubuntu 22.04 LTS (2022). https://doi.org/10.5281/zenodo.7113223

85. Filieri, A., Tamburrelli, G., Ghezzi, C.: Supporting self-adaptation via quantitative verification and sensitivity analysis at run time. IEEE Trans. Software Eng. **42**(1), 75–99 (2016). https://doi.org/10.1109/TSE.2015.2421318

86. Fontanarrosa, P., Doosthosseini, H., Borujeni, A.E., Dorfan, Y., Voigt, C.A., Myers, C.: Genetic circuit dynamics: hazard and glitch analysis. ACS Synth. Biol. **9**(9), 2324–2338 (2020). https://doi.org/10.1021/acssynbio.0c00055

87. Forejt, V., Kwiatkowska, M., Norman, G., Parker, D., Qu, H.: Quantitative multi-objective verification for probabilistic systems. In: Abdulla, P.A., Leino, K.R.M. (eds.) TACAS 2011. LNCS, vol. 6605, pp. 112–127. Springer, Heidelberg (2011). https://doi.org/10.1007/978-3-642-19835-9_11

88. Forejt, V., Kwiatkowska, M., Parker, D.: Pareto curves for probabilistic model checking. In: Chakraborty, S., Mukund, M. (eds.) ATVA 2012. LNCS, pp. 317–332. Springer, Heidelberg (2012). https://doi.org/10.1007/978-3-642-33386-6_25

89. Fränzle, M., Hahn, E.M., Hermanns, H., Wolovick, N., Zhang, L.: Measurability and safety verification for stochastic hybrid systems. In: Caccamo, M., Frazzoli, E., Grosu, R. (eds.) 14th ACM International Conference on Hybrid Systems: Computation and Control (HSCC), pp. 43–52. ACM (2011). https://doi.org/10.1145/1967701.1967710

90. Frehse, G., Althoff, M. (eds.): 4th International Workshop on Applied Verification of Continuous and Hybrid Systems (ARCH), EPiC Series in Computing, vol. 48. EasyChair (2017). https://easychair.org/publications/volume/ARCH17

91. Fu, C., Hahn, E.M., Li, Y., Schewe, S., Sun, M., Turrini, A., Zhang, L.: EPMC gets knowledge in multi-agent systems. In: Finkbeiner, B., Wies, T. (eds.) VMCAI 2022. LNCS, vol. 13182, pp. 93–107. Springer, Cham (2022). https://doi.org/10.1007/978-3-030-94583-1_5

92. Gainer, P., Hahn, E.M., Schewe, S.: Accelerated model checking of parametric Markov chains. In: Lahiri, S.K., Wang, C. (eds.) ATVA 2018. LNCS, vol. 11138, pp. 300–316. Springer, Cham (2018). https://doi.org/10.1007/978-3-030-01090-4_18

93. Gillespie, D.T.: Exact stochastic simulation of coupled chemical reactions. J. Phys. Chem. **81**(25), 2340–2361 (1977). https://doi.org/10.1021/j100540a008

94. Goldberg, F., Vesely, W.: Fault Tree Handbook. NUREG-0492, Systems and Reliability Research, Office of Nuclear Regulatory Research, U.S. Nuclear Regulatory Commission (1981)

95. Goutsias, J.: Quasiequilibrium approximation of fast reaction kinetics in stochastic biochemical systems. J. Chem. Phys. **122**(18) (2005). https://doi.org/10.1063/1.1889434

96. Goyal, V., Grand-Clément, J.: Robust Markov decision processes: beyond rectangularity. Math. Oper. Res. **48**(1), 203–226 (2023). https://doi.org/10.1287/moor.2022.1259

97. Gros, T.P., Hermanns, H., Hoffmann, J., Klauck, M., Köhl, M.A., Wolf, V.: MoGym: using formal models for training and verifying decision-making agents. In: Shoham, S., Vizel, Y. (eds.) CAV 2022. LNCS, vol. 13372, pp. 430–443. Springer, Cham (2022). https://doi.org/10.1007/978-3-031-13188-2_21

98. Gros, T.P., Hermanns, H., Hoffmann, J., Klauck, M., Steinmetz, M.: Deep statistical model checking. In: Gotsman, A., Sokolova, A. (eds.) FORTE 2020. LNCS, vol. 12136, pp. 96–114. Springer, Cham (2020). https://doi.org/10.1007/978-3-030-50086-3_6

99. Gross, D., Jansen, N., Junges, S., Pérez, G.A.: COOL-MC: a comprehensive tool for reinforcement learning and model checking. In: Dong, W., Talpin, J.P. (eds.) SETTA 2022. LNCS, vol. 13649, pp. 41–49. Springer, Cham (2022). https://doi.org/10.1007/978-3-031-21213-0_3

100. Grover, K.: QComp LRA results (2023). https://doi.org/10.5281/zenodo.8219191

101. Guck, D., Timmer, M., Hatefi, H., Ruijters, E., Stoelinga, M.: Modelling and analysis of Markov reward automata. In: Cassez, F., Raskin, J.-F. (eds.) ATVA 2014. LNCS, vol. 8837, pp. 168–184. Springer, Cham (2014). https://doi.org/10.1007/978-3-319-11936-6_13

102. Hahn, E.M., Hartmanns, A.: Symblicit exploration and elimination for probabilistic model checking. In: Hung, C.C., Hong, J., Bechini, A., Song, E. (eds.) 36th ACM/SIGAPP Symposium on Applied Computing (SAC), pp. 1798–1806. ACM (2021). https://doi.org/10.1145/3412841.3442052

103. Hahn, E.M., et al.: The 2019 comparison of tools for the analysis of quantitative formal models. In: Beyer, D., Huisman, M., Kordon, F., Steffen, B. (eds.) TACAS 2019. LNCS, vol. 11429, pp. 69–92. Springer, Cham (2019). https://doi.org/10.1007/978-3-030-17502-3_5

104. Hahn, E.M., Hartmanns, A., Hermanns, H.: Reachability and reward checking for stochastic timed automata. Electron. Commun. Eur. Assoc. Softw. Sci. Technol. **70** (2014). https://doi.org/10.14279/tuj.eceasst.70.968

105. Hahn, E.M., Hartmanns, A., Hermanns, H., Katoen, J.P.: A compositional modelling and analysis framework for stochastic hybrid systems. Formal Methods Syst. Des. **43**(2), 191–232 (2013). https://doi.org/10.1007/s10703-012-0167-z

106. Hahn, E.M., Hermanns, H., Wachter, B., Zhang, L.: INFAMY: an infinite-state Markov model checker. In: Bouajjani, A., Maler, O. (eds.) CAV 2009. LNCS, vol. 5643, pp. 641–647. Springer, Heidelberg (2009). https://doi.org/10.1007/978-3-642-02658-4_49

107. Hahn, E.M., Hermanns, H., Zhang, L.: Probabilistic reachability for parametric Markov models. Int. J. Softw. Tools Technol. Transf. **13**(1), 3–19 (2011). https://doi.org/10.1007/s10009-010-0146-x

108. Hahn, E.M., Li, G., Schewe, S., Turrini, A., Zhang, L.: Lazy probabilistic model checking without determinisation. In: Aceto, L., de Frutos-Escrig, D. (eds.) 26th International Conference on Concurrency Theory (CONCUR). LIPIcs, vol. 42, pp. 354–367. Schloss Dagstuhl – Leibniz-Zentrum für Informatik (2015). https://doi.org/10.4230/LIPIcs.CONCUR.2015.354

109. Hahn, E.M., Li, Y., Schewe, S., Turrini, A., Zhang, L.: iscasMc: a web-based probabilistic model checker. In: Jones, C., Pihlajasaari, P., Sun, J. (eds.) FM 2014. LNCS, vol. 8442, pp. 312–317. Springer, Cham (2014). https://doi.org/10.1007/978-3-319-06410-9_22

110. Hahn, E.M., Perez, M., Schewe, S., Somenzi, F., Trivedi, A., Wojtczak, D.: Good-for-MDPs automata for probabilistic analysis and reinforcement learning. In: TACAS 2020. LNCS, vol. 12078, pp. 306–323. Springer, Cham (2020). https://doi.org/10.1007/978-3-030-45190-5_17

111. Hartmanns, A.: Correct probabilistic model checking with floating-point arithmetic. In: Fisman, D., Rosu, G. (eds.) TACAS 2022. LNCS, vol. 13244, pp. 41–59. Springer, Cham (2022). https://doi.org/10.1007/978-3-030-99527-0_3

112. Hartmanns, A., Hermanns, H.: A Modest approach to checking probabilistic timed automata. In: 6th International Conference on the Quantitative Evaluation of Systems (QEST), pp. 187–196. IEEE Computer Society (2009). https://doi.org/10.1109/QEST.2009.41

113. Hartmanns, A., Hermanns, H.: The modest toolset: an integrated environment for quantitative modelling and verification. In: Ábrahám, E., Havelund, K. (eds.) TACAS 2014. LNCS, vol. 8413, pp. 593–598. Springer, Heidelberg (2014). https://doi.org/10.1007/978-3-642-54862-8_51

114. Hartmanns, A., Hermanns, H.: Explicit model checking of very large MDP using partitioning and secondary storage. In: Finkbeiner, B., Pu, G., Zhang, L. (eds.) ATVA 2015. LNCS, vol. 9364, pp. 131–147. Springer, Cham (2015). https://doi.org/10.1007/978-3-319-24953-7_10

115. Hartmanns, A., Junges, S., Katoen, J.P., Quatmann, T.: Multi-cost bounded tradeoff analysis in MDP. J. Autom. Reason. **64**(7), 1483–1522 (2020). https://doi.org/10.1007/s10817-020-09574-9

116. Hartmanns, A., Junges, S., Quatmann, T., Weininger, M.: A practitioner's guide to MDP model checking algorithms. In: Sankaranarayanan, S., Sharygina, N.

(eds.) TACAS 2023. LNCS, vol. 13993, pp. 469–488. Springer, Cham (2023). https://doi.org/10.1007/978-3-031-30823-9_24

117. Hartmanns, A., Kaminski, B.L.: Optimistic value iteration. In: Lahiri, S.K., Wang, C. (eds.) CAV 2020. LNCS, vol. 12225, pp. 488–511. Springer, Cham (2020). https://doi.org/10.1007/978-3-030-53291-8_26

118. Hartmanns, A., Katoen, J.-P., Kohlen, B., Spel, J.: Tweaking the odds in probabilistic timed automata. In: Abate, A., Marin, A. (eds.) QEST 2021. LNCS, vol. 12846, pp. 39–58. Springer, Cham (2021). https://doi.org/10.1007/978-3-030-85172-9_3

119. Hartmanns, A., Klauck, M., Parker, D., Quatmann, T., Ruijters, E.: The quantitative verification benchmark set. In: Vojnar, T., Zhang, L. (eds.) TACAS 2019. LNCS, vol. 11427, pp. 344–350. Springer, Cham (2019). https://doi.org/10.1007/978-3-030-17462-0_20

120. Hartmanns, A., Sedwards, S., D'Argenio, P.R.: Efficient simulation-based verification of probabilistic timed automata. In: 2017 Winter Simulation Conference (WSC), pp. 1419–1430. IEEE (2017). https://doi.org/10.1109/WSC.2017.8247885

121. Hasenauer, J., Wolf, V., Kazeroonian, A., Theis, F.J.: Method of conditional moments (MCM) for the chemical master equation. J. Math. Biol. **69**(3), 687–735 (2013). https://doi.org/10.1007/s00285-013-0711-5

122. Heck, L., Spel, J., Junges, S., Moerman, J., Katoen, J.P.: Gradient-descent for randomized controllers under partial observability. In: Finkbeiner, B., Wies, T. (eds.) VMCAI 2022. LNCS, vol. 13182, pp. 127–150. Springer, Cham (2022). https://doi.org/10.1007/978-3-030-94583-1_7

123. Heidelberger, P.: Fast simulation of rare events in queueing and reliability models. In: Donatiello, L., Nelson, R. (eds.) Performance/SIGMETRICS -1993. LNCS, vol. 729, pp. 165–202. Springer, Heidelberg (1993). https://doi.org/10.1007/BFb0013853

124. Helfrich, M., Ceska, M., Kretínský, J., Marticek, S.: Abstraction-based segmental simulation of chemical reaction networks. In: Petre, I., Paun, A. (eds.) CMSB 2022. LNCS, vol. 13447, pp. 41–60. Springer, Cham (2022). https://doi.org/10.1007/978-3-031-15034-0_3

125. Hensel, C., Junges, S., Katoen, J.P., Quatmann, T., Volk, M.: The probabilistic model checker Storm. Int. J. Softw. Tools Technol. Transf. **24**(4), 589–610 (2022). https://doi.org/10.1007/s10009-021-00633-z

126. Henzinger, T.A., Mikeev, L., Mateescu, M., Wolf, V.: Hybrid numerical solution of the chemical master equation. In: Quaglia, P. (ed.) 8th International Conference on Computational Methods in Systems Biology (CMSB), pp. 55–65. ACM (2010). https://doi.org/10.1145/1839764.1839772

127. Hermanns, H., Meyer-Kayser, J., Siegle, M.: Multi terminal binary decision diagrams to represent and analyse continuous time Markov chains. In: Plateau, B., Stewart, W., Silva, M. (eds.) 3rd International Workshop on Numerical Solution of Markov Chains (NSMC), pp. 188–207. Prensas Universitarias de Zaragoza (1999)

128. Holtzen, S., Junges, S., Vazquez-Chanlatte, M., Millstein, T., Seshia, S.A., Van den Broeck, G.: Model checking finite-horizon Markov chains with probabilistic inference. In: Silva, A., Leino, K.R.M. (eds.) CAV 2021. LNCS, vol. 12760, pp. 577–601. Springer, Cham (2021). https://doi.org/10.1007/978-3-030-81688-9_27

129. Israelsen, B., Taylor, L., Zhang, Z.: Efficient trace generation for rare-event analysis in chemical reaction networks. In: Caltais, G., Schilling, C. (eds.) SPIN 2023. LNCS, vol. 13872, pp. 83–102. Springer, Cham (2023). https://doi.org/10.1007/978-3-031-32157-3_5

130. Itoh, H., Nakamura, K.: Partially observable Markov decision processes with imprecise parameters. Artif. Intell. **171**(8–9), 453–490 (2007). https://doi.org/10.1016/j.artint.2007.03.004

131. Jackson, J.R.: Networks of waiting lines. Oper. Res. **5**, 518–521 (1957)

132. Jansen, N., Junges, S., Katoen, J.P.: Parameter synthesis in Markov models: a gentle survey. In: Raskin, J.F., Chatterjee, K., Doyen, L., Majumdar, R. (eds.) Principles of Systems Design. LNCS, vol. 13660, pp. 407–437. Springer, Cham (2022). https://doi.org/10.1007/978-3-031-22337-2_20

133. Jégourel, C., Legay, A., Sedwards, S.: Command-based importance sampling for statistical model checking. Theor. Comput. Sci. **649**, 1–24 (2016). https://doi.org/10.1016/j.tcs.2016.08.009

134. Jeppson, J., et al.: STAMINA in C++: modernizing an infinite-state probabilistic model checker. In: Jansen, N., Tribastone, M. (eds.) QEST 2023. LNCS, vol. 14287, pp. 101–109. Springer, Cham (2023). https://doi.org/10.1007/978-3-031-43835-6_7

135. John, T., Jantsch, S., Baier, C., Klüppelholz, S.: From Emerson-Lei automata to deterministic, limit-deterministic or good-for-MDP automata. Innov. Syst. Softw. Eng. **18**(3), 385–403 (2022). https://doi.org/10.1007/s11334-022-00445-7

136. Junges, S.: Parameter synthesis in Markov models. Ph.D. thesis, RWTH Aachen University (2020). https://publications.rwth-aachen.de/record/783179

137. Junges, S.: sjunges/parametric-Markov-models: 0.2 (2023). https://doi.org/10.5281/zenodo.10646479

138. Junges, S., et al.: Parameter synthesis for Markov models. CoRR **abs/1903.07993** (2019). https://doi.org/10.48550/arXiv.1903.07993

139. Junges, S., Jansen, N., Seshia, S.A.: Enforcing almost-sure reachability in POMDPs. In: Silva, A., Leino, K.R.M. (eds.) CAV 2021. LNCS, vol. 12760, pp. 602–625. Springer, Cham (2021). https://doi.org/10.1007/978-3-030-81688-9_28

140. Junges, S., Spaan, M.T.J.: Abstraction-refinement for hierarchical probabilistic models. In: Shoham, S., Vizel, Y. (eds.) CAV 2022. LNCS, vol. 13371, pp. 102–123. Springer, Cham (2022)

141. Kahn, H., Harris, T.E.: Estimation of particle transmission by random sampling. Natl. Bureau Stand. Appl. Math. Lett. **12**, 27–30 (1951)

142. Klauck, M., Steinmetz, M., Hoffmann, J., Hermanns, H.: Bridging the gap between probabilistic model checking and probabilistic planning: survey, compilations, and empirical comparison. J. Artif. Intell. Res. **68**, 247–310 (2020). https://doi.org/10.1613/jair.1.11595

143. Kochenderfer, M.J.: Decision Making Under Uncertainty: Theory and Application. MIT Press, Cambridge (2015)

144. Köhl, M.A.: QComp 2023: State space exploration artifact (2024). https://doi.org/10.5281/zenodo.10626177

145. Köhl, M.A., Klauck, M., Hermanns, H.: Momba: JANI meets Python. In: TACAS 2021. LNCS, vol. 12652, pp. 389–398. Springer, Cham (2021). https://doi.org/10.1007/978-3-030-72013-1_23

146. Kretínský, J.: LTL-constrained steady-state policy synthesis. In: Zhou, Z.H. (ed.) 30th International Joint Conference on Artificial Intelligence (IJCAI), pp. 4104–4111. ijcai.org (2021). https://doi.org/10.24963/ijcai.2021/565

147. Křetínský, J., Esparza, J.: Deterministic automata for the (F,G)-fragment of LTL. In: Madhusudan, P., Seshia, S.A. (eds.) CAV 2012. LNCS, vol. 7358, pp. 7–22. Springer, Heidelberg (2012). https://doi.org/10.1007/978-3-642-31424-7_7

148. Kretínský, J., Meggendorfer, T.: Of cores: a partial-exploration framework for Markov decision processes. Log. Methods Comput. Sci. **16**(4) (2020). https://doi.org/10.23638/LMCS-16(4:3)2020

149. Kretínský, J., Meggendorfer, T., Sickert, S.: LTL Store: Repository of LTL formulae from literature and case studies. CoRR **abs/1807.03296** (2018). https://doi.org/10.48550/arXiv.1807.03296

150. Křetínský, J., Meggendorfer, T., Sickert, S.: Owl: a library for ω-words, automata, and LTL. In: Lahiri, S.K., Wang, C. (eds.) ATVA 2018. LNCS, vol. 11138, pp. 543–550. Springer, Cham (2018). https://doi.org/10.1007/978-3-030-01090-4_34

151. Křetínský, J., Meggendorfer, T., Sickert, S., Ziegler, C.: Rabinizer 4: from LTL to your favourite deterministic automaton. In: Chockler, H., Weissenbacher, G. (eds.) CAV 2018. LNCS, vol. 10981, pp. 567–577. Springer, Cham (2018). https://doi.org/10.1007/978-3-319-96145-3_30

152. Kretínský, J., Meggendorfer, T., Weininger, M.: Stopping criteria for value iteration on stochastic games with quantitative objectives. In: 38th Annual ACM/IEEE Symposium on Logic in Computer Science (LICS), pp. 1–14 (2023). https://doi.org/10.1109/LICS56636.2023.10175771

153. Kretínský, J., Michel, F., Michel, L., Pérez, G.A.: Finite-memory near-optimal learning for Markov decision processes with long-run average reward. In: Adams, R.P., Gogate, V. (eds.) 36th Conference on Uncertainty in Artificial Intelligence (UAI). Proceedings of Machine Learning Research, vol. 124, pp. 1149–1158. AUAI Press (2020)

154. Kretínský, J., Ramneantu, E., Slivinskiy, A., Weininger, M.: Comparison of algorithms for simple stochastic games. Inf. Comput. **289**(Part), 104885 (2022). https://doi.org/10.1016/j.ic.2022.104885

155. Kurniawati, H., Hsu, D., Lee, W.S.: SARSOP: efficient point-based POMDP planning by approximating optimally reachable belief spaces. In: Brock, O., Trinkle, J., Ramos, F. (eds.) Robotics: Science and Systems IV. The MIT Press (2008). https://doi.org/10.15607/RSS.2008.IV.009

156. Kurose, J.F., Ross, K.W.: Computer Networking - A Top-down Approach Featuring the Internet. Addison-Wesley-Longman, Boston (2001)

157. Kuwahara, H., Mura, I.: An efficient and exact stochastic simulation method to analyze rare events in biochemical systems. J. Chem. Phys. **129**(16) (2008). https://doi.org/10.1063/1.2987701

158. Kwiatkowska, M., Norman, G., Parker, D., Santos, G.: PRISM-games 3.0: stochastic game verification with concurrency, equilibria and time. In: Lahiri, S.K., Wang, C. (eds.) CAV 2020. LNCS, vol. 12225, pp. 475–487. Springer, Cham (2020). https://doi.org/10.1007/978-3-030-53291-8_25

159. Kwiatkowska, M., Norman, G., Parker, D., Santos, G.: Automatic verification of concurrent stochastic systems. Formal Methods Syst. Des. **58**(1–2), 188–250 (2021). https://doi.org/10.1007/s10703-020-00356-y

160. Kwiatkowska, M., Norman, G., Parker, D., Santos, G.: Correlated equilibria and fairness in concurrent stochastic games. In: Fisman, D., Rosu, G. (eds.) TACAS 2022. LNCS, vol. 13244, pp. 60–78. Springer, Cham (2022). https://doi.org/10.1007/978-3-030-99527-0_4

161. Kwiatkowska, M., Norman, G., Parker, D., Santos, G.: Symbolic verification and strategy synthesis for turn-based stochastic games. In: Raskin, J.F., Chatterjee, K., Doyen, L., Majumdar, R. (eds.) Principles of Systems Design. LNCS, vol. 13660, pp. 388–406. Springer, Cham (2022). https://doi.org/10.1007/978-3-031-22337-2_19

162. Kwiatkowska, M., Norman, G., Parker, D.: Stochastic games for verification of probabilistic timed automata. In: Ouaknine, J., Vaandrager, F.W. (eds.) FORMATS 2009. LNCS, vol. 5813, pp. 212–227. Springer, Heidelberg (2009). https://doi.org/10.1007/978-3-642-04368-0_17

163. Kwiatkowska, M., Norman, G., Parker, D.: PRISM 4.0: verification of probabilistic real-time systems. In: Gopalakrishnan, G., Qadeer, S. (eds.) CAV 2011. LNCS, vol. 6806, pp. 585–591. Springer, Heidelberg (2011). https://doi.org/10.1007/978-3-642-22110-1_47

164. Kwiatkowska, M.Z., Norman, G., Parker, D.: The PRISM benchmark suite. In: 9th International Conference on the Quantitative Evaluation of Systems (QEST), pp. 203–204. IEEE Computer Society (2012). https://doi.org/10.1109/QEST.2012.14

165. Kwiatkowska, M.Z., Norman, G., Parker, D., Sproston, J.: Performance analysis of probabilistic timed automata using digital clocks. Formal Methods Syst. Des. **29**(1), 33–78 (2006). https://doi.org/10.1007/s10703-006-0005-2

166. Kwiatkowska, M.Z., Norman, G., Segala, R., Sproston, J.: Automatic verification of real-time systems with discrete probability distributions. Theor. Comput. Sci. **282**(1), 101–150 (2002). https://doi.org/10.1016/S0304-3975(01)00046-9

167. Kwiatkowska, M.Z., Norman, G., Sproston, J., Wang, F.: Symbolic model checking for probabilistic timed automata. Inf. Comput. **205**(7), 1027–1077 (2007). https://doi.org/10.1016/j.ic.2007.01.004

168. Lanotte, R., Maggiolo-Schettini, A., Troina, A.: Parametric probabilistic transition systems for system design and analysis. Formal Aspects Comput. **19**(1), 93–109 (2007). https://doi.org/10.1007/s00165-006-0015-2

169. Legay, A., Lukina, A., Traonouez, L.M., Yang, J., Smolka, S.A., Grosu, R.: Statistical model checking. In: Steffen, B., Woeginger, G. (eds.) Computing and Software Science. LNCS, vol. 10000, pp. 478–504. Springer, Cham (2019). https://doi.org/10.1007/978-3-319-91908-9_23

170. Li, M., Turrini, A., Hahn, E.M., She, Z., Zhang, L.: Probabilistic preference planning problem for Markov decision processes. IEEE Trans. Software Eng. **48**(5), 1545–1559 (2022). https://doi.org/10.1109/TSE.2020.3024215

171. Lovejoy, W.S.: Computationally feasible bounds for partially observed Markov decision processes. Oper. Res. **39**(1), 162–175 (1991). https://doi.org/10.1287/opre.39.1.162

172. Madani, O., Hanks, S., Condon, A.: On the undecidability of probabilistic planning and related stochastic optimization problems. Artif. Intell. **147**(1–2), 5–34 (2003). https://doi.org/10.1016/S0004-3702(02)00378-8

173. Madsen, C., Zhang, Z., Roehner, N., Winstead, C., Myers, C.J.: Stochastic model checking of genetic circuits. ACM J. Emerg. Technol. Comput. Syst. **11**(3), 23:1–23:21 (2014). https://doi.org/10.1145/2644817

174. Major, J., Blahoudek, F., Strejček, J., Sasaráková, M., Zbončáková, T.: `ltl3tela`: LTL to small deterministic or nondeterministic Emerson-Lei automata. In: Chen, Y.-F., Cheng, C.-H., Esparza, J. (eds.) ATVA 2019. LNCS, vol. 11781, pp. 357–365. Springer, Cham (2019). https://doi.org/10.1007/978-3-030-31784-3_21

175. Mannor, S., Simester, D., Sun, P., Tsitsiklis, J.N.: Bias and variance approximation in value function estimates. Manag. Sci. **53**(2), 308–322 (2007). https://doi.org/10.1287/mnsc.1060.0614

176. Mausam, Kolobov, A.: Planning with Markov Decision Processes: An AI Perspective. Synthesis Lectures on Artificial Intelligence and Machine Learning, Morgan & Claypool Publishers (2012). https://doi.org/10.2200/S00426ED1V01Y201206AIM017

177. McMillan, K.L., Zuck, L.D.: Compositional testing of Internet protocols. In: 2019 IEEE Secure Development Conference (SecDev), pp. 161–174. IEEE (2019). https://doi.org/10.1109/SecDev.2019.00031
178. Mediouni, B.L., Nouri, A., Bozga, M., Dellabani, M., Legay, A., Bensalem, S.: SBIP 2.0: statistical model checking stochastic real-time systems. In: Lahiri, S.K., Wang, C. (eds.) ATVA 2018. LNCS, vol. 11138, pp. 536–542. Springer, Cham (2018). https://doi.org/10.1007/978-3-030-01090-4_33
179. Meggendorfer, T.: PET - a partial exploration tool for probabilistic verification. In: Bouajjani, A., Holík, L., Wu, Z. (eds.) ATVA 2022. LNCS, vol. 13505, pp. 320–326. Springer, Cham (2022). https://doi.org/10.1007/978-3-031-19992-9_20
180. Meggendorfer, T.: QComp 2023: Stochastic games – evaluation (2023). https://doi.org/10.5281/zenodo.7831387
181. Müller, D., Sickert, S.: LTL to deterministic Emerson-Lei automata. In: Bouyer, P., Orlandini, A., Pietro, P.S. (eds.) 8th International Symposium on Games, Automata, Logics and Formal Verification (GandALF). EPTCS, vol. 256, pp. 180–194 (2017). https://doi.org/10.4204/EPTCS.256.13
182. Munsky, B., Khammash, M.: The finite state projection algorithm for the solution of the chemical master equation. J. Chem. Phys. **124**(4) (2006). https://doi.org/10.1063/1.2145882
183. Neupane, T., Myers, C.J., Madsen, C., Zheng, H., Zhang, Z.: STAMINA: STochastic approximate model-checker for INfinite-state analysis. In: Dillig, I., Tasiran, S. (eds.) CAV 2019. LNCS, vol. 11561, pp. 540–549. Springer, Cham (2019). https://doi.org/10.1007/978-3-030-25540-4_31
184. Neupane, T., Zhang, Z., Madsen, C., Zheng, H., Myers, C.J.: Approximation techniques for stochastic analysis of biological systems. In: Liò, P., Zuliani, P. (eds.) Automated Reasoning for Systems Biology and Medicine. CB, vol. 30, pp. 327–348. Springer, Cham (2019). https://doi.org/10.1007/978-3-030-17297-8_12
185. Nicola, V.F., Shahabuddin, P., Nakayama, M.K.: Techniques for fast simulation of models of highly dependable systems. IEEE Trans. Reliab. **50**(3), 246–264 (2001). https://doi.org/10.1109/24.974122
186. Niehage, M., Hartmanns, A., Remke, A.: Learning optimal decisions for stochastic hybrid systems. In: Arun-Kumar, S., Méry, D., Saha, I., Zhang, L. (eds.) 19th ACM-IEEE International Conference on Formal Methods and Models for System Design (MEMOCODE), pp. 44–55. ACM (2021). https://doi.org/10.1145/3487212.3487339
187. Nilim, A., Ghaoui, L.E.: Robust control of Markov decision processes with uncertain transition matrices. Oper. Res. **53**(5), 780–798 (2005). https://doi.org/10.1287/opre.1050.0216
188. Norman, G., Parker, D., Zou, X.: Verification and control of partially observable probabilistic systems. Real Time Syst. **53**(3), 354–402 (2017). https://doi.org/10.1007/s11241-017-9269-4
189. Pai, G.J., Dugan, J.B.: Automatic synthesis of dynamic fault trees from UML system models. In: 13th International Symposium on Software Reliability Engineering (ISSRE), pp. 243–256. IEEE Computer Society (2002). https://doi.org/10.1109/ISSRE.2002.1173261
190. Phalakarn, K., Takisaka, T., Haas, T., Hasuo, I.: Widest paths and global propagation in bounded value iteration for stochastic games. In: Lahiri, S.K., Wang, C. (eds.) CAV 2020. LNCS, vol. 12225, pp. 349–371. Springer, Cham (2020). https://doi.org/10.1007/978-3-030-53291-8_19

191. Pnueli, A.: The temporal logic of programs. In: 18th Annual Symposium on Foundations of Computer Science (FOCS), pp. 46–57. IEEE Computer Society (1977). https://doi.org/10.1109/SFCS.1977.32

192. Pranger, S., Könighofer, B., Posch, L., Bloem, R.: TEMPEST - synthesis tool for reactive systems and shields in probabilistic environments. In: Hou, Z., Ganesh, V. (eds.) ATVA 2021. LNCS, vol. 12971, pp. 222–228. Springer, Cham (2021). https://doi.org/10.1007/978-3-030-88885-5_15

193. Puggelli, A., Li, W., Sangiovanni-Vincentelli, A.L., Seshia, S.A.: Polynomial-time verification of PCTL properties of MDPs with convex uncertainties. In: Sharygina, N., Veith, H. (eds.) CAV 2013. LNCS, vol. 8044, pp. 527–542. Springer, Heidelberg (2013). https://doi.org/10.1007/978-3-642-39799-8_35

194. Puterman, M.L.: Markov Decision Processes: Discrete Stochastic Dynamic Programming. Wiley Series in Probability and Statistics. Wiley (1994). https://doi.org/10.1002/9780470316887

195. Quatmann, T.: Replication package: QComp 2023 – multi-objective analysis (2023). https://doi.org/10.5281/zenodo.8063883

196. Quatmann, T., Junges, S., Katoen, J.P.: Markov automata with multiple objectives. Formal Methods Syst. Des. **60**(1), 33–86 (2022). https://doi.org/10.1007/s10703-021-00364-6

197. Quatmann, T., Katoen, J.-P.: Multi-objective optimization of long-run average and total rewards. In: TACAS 2021. LNCS, vol. 12651, pp. 230–249. Springer, Cham (2021). https://doi.org/10.1007/978-3-030-72016-2_13

198. Reijsbergen, D., de Boer, P.T., Scheinhardt, W.R.W., Juneja, S.: Path-ZVA: general, efficient, and automated importance sampling for highly reliable Markovian systems. ACM Trans. Model. Comput. Simul. **28**(3), 22:1–22:25 (2018). https://doi.org/10.1145/3161569

199. Roberts, R., Neupane, T., Buecherl, L., Myers, C.J., Zhang, Z.: STAMINA 2.0: improving scalability of infinite-state stochastic model checking. In: Finkbeiner, B., Wies, T. (eds.) VMCAI 2022. LNCS, vol. 13182, pp. 319–331. Springer, Cham (2022). https://doi.org/10.1007/978-3-030-94583-1_16

200. Ruijters, E., Reijsbergen, D., de Boer, P.T., Stoelinga, M.: Rare event simulation for dynamic fault trees. Reliab. Eng. Syst. Saf. **186**, 220–231 (2019). https://doi.org/10.1016/j.ress.2019.02.004

201. Russell, S., Norvig, P.: Artificial Intelligence: A Modern Approach, 4th edn. Pearson, London (2020)

202. Salmani, B., Katoen, J.-P.: Fine-tuning the odds in Bayesian networks. In: Vejnarová, J., Wilson, N. (eds.) ECSQARU 2021. LNCS (LNAI), vol. 12897, pp. 268–283. Springer, Cham (2021). https://doi.org/10.1007/978-3-030-86772-0_20

203. Schwartz, A.: A reinforcement learning method for maximizing undiscounted rewards. In: Utgoff, P.E. (ed.) 10th International Conference on Machine Learning (ICML), pp. 298–305. Morgan Kaufmann (1993). https://doi.org/10.1016/b978-1-55860-307-3.50045-9

204. Shani, G., Pineau, J., Kaplow, R.: A survey of point-based POMDP solvers. Auton. Agents Multi Agent Syst. **27**(1), 1–51 (2013). https://doi.org/10.1007/s10458-012-9200-2

205. Sickert, S., Esparza, J., Jaax, S., Křetínský, J.: Limit-deterministic Büchi automata for linear temporal logic. In: Chaudhuri, S., Farzan, A. (eds.) CAV 2016. LNCS, vol. 9780, pp. 312–332. Springer, Cham (2016). https://doi.org/10.1007/978-3-319-41540-6_17

206. Sickert, S., Křetínský, J.: MoChiBA: probabilistic LTL model checking using limit-deterministic Büchi automata. In: Artho, C., Legay, A., Peled, D. (eds.) ATVA 2016. LNCS, vol. 9938, pp. 130–137. Springer, Cham (2016). https://doi.org/10.1007/978-3-319-46520-3_9

207. Spel, J., Junges, S., Katoen, J.-P.: Are parametric Markov chains monotonic? In: Chen, Y.-F., Cheng, C.-H., Esparza, J. (eds.) ATVA 2019. LNCS, vol. 11781, pp. 479–496. Springer, Cham (2019). https://doi.org/10.1007/978-3-030-31784-3_28

208. Spel, J., Junges, S., Katoen, J.-P.: Finding provably optimal Markov chains. In: TACAS 2021. LNCS, vol. 12651, pp. 173–190. Springer, Cham (2021). https://doi.org/10.1007/978-3-030-72016-2_10

209. Suilen, M., Jansen, N., Cubuktepe, M., Topcu, U.: Robust policy synthesis for uncertain POMDPs via convex optimization. In: Bessiere, C. (ed.) 29th International Joint Conference on Artificial Intelligence (IJCAI), pp. 4113–4120. ijcai.org (2020). https://doi.org/10.24963/ijcai.2020/569

210. Suilen, M., Simão, T.D., Parker, D., Jansen, N.: Robust anytime learning of Markov decision processes. In: NeurIPS (2022)

211. Taylor, L., Israelsen, B., Zhang, Z.: Cycle and commute: rare-event probability verification for chemical reaction networks. In: Nadel, A., Rozier, K.Y. (eds.) 23rd Conference on Formal Methods in Computer-Aided Design (FMCAD), pp. 284–293. TU Wien Academic Press (2023). https://doi.org/10.34727/2023/ISBN.978-3-85448-060-0_37

212. Van Kampen, N.G.: Stochastic Processes in Physics and Chemistry, vol. 1. Elsevier, Amsterdam (1992)

213. Vardi, M.Y., Wolper, P.: An automata-theoretic approach to automatic program verification (preliminary report). In: 1st Annual IEEE Symposium on Logic in Computer Science (LICS), pp. 332–344. IEEE Computer Society (1986)

214. Velasquez, A., Alkhouri, I., Beckus, A., Trivedi, A., Atia, G.K.: Controller synthesis for omega-regular and steady-state specifications. In: Faliszewski, P., Mascardi, V., Pelachaud, C., Taylor, M.E. (eds.) 21st International Conference on Autonomous Agents and Multiagent Systems (AAMAS), pp. 1310–1318. International Foundation for Autonomous Agents and Multiagent Systems (2022). https://doi.org/10.5555/3535850.3535996

215. Villén-Altamirano, J.: RESTART vs splitting: a comparative study. Perform. Evaluation **121–122**, 38–47 (2018). https://doi.org/10.1016/j.peva.2018.02.002

216. Volk, M., Junges, S., Katoen, J.-P.: Advancing dynamic fault tree analysis - get succinct state spaces fast and synthesise failure rates. In: Skavhaug, A., Guiochet, J., Bitsch, F. (eds.) SAFECOMP 2016. LNCS, vol. 9922, pp. 253–265. Springer, Cham (2016). https://doi.org/10.1007/978-3-319-45477-1_20

217. Volk, M., Junges, S., Katoen, J.P.: Fast dynamic fault tree analysis by model checking techniques. IEEE Trans. Ind. Informatics **14**(1), 370–379 (2018). https://doi.org/10.1109/TII.2017.2710316

218. Wiesemann, W., Kuhn, D., Sim, M.: Distributionally robust convex optimization. Oper. Res. **62**(6), 1358–1376 (2014). https://doi.org/10.1287/opre.2014.1314

219. Winkler, T., Junges, S., Pérez, G.A., Katoen, J.P.: On the complexity of reachability in parametric Markov decision processes. In: Fokkink, W.J., van Glabbeek, R. (eds.) 30th International Conference on Concurrency Theory (CONCUR). LIPIcs, vol. 140, pp. 14:1–14:17. Schloss Dagstuhl - Leibniz-Zentrum für Informatik (2019). https://doi.org/10.4230/LIPIcs.CONCUR.2019.14

220. Wolff, E.M., Topcu, U., Murray, R.M.: Robust control of uncertain Markov decision processes with temporal logic specifications. In: 51th IEEE Conference on

Decision and Control (CDC), pp. 3372–3379. IEEE (2012). https://doi.org/10.1109/CDC.2012.6426174

221. Younes, H.L.S., Simmons, R.G.: Probabilistic verification of discrete event systems using acceptance sampling. In: Brinksma, E., Larsen, K.G. (eds.) CAV 2002. LNCS, vol. 2404, pp. 223–235. Springer, Heidelberg (2002). https://doi.org/10.1007/3-540-45657-0_17

222. Yu, H., Bertsekas, D.P.: Discretized approximations for POMDP with average cost. In: Chickering, D.M., Halpern, J.Y. (eds.) 20th Conference on Uncertainty in Artificial Intelligence (UAI), p. 519. AUAI Press (2004)

223. Zhang, J., Watson, L.T., Cao, Y.: Adaptive aggregation method for the chemical master equation. Int. J. Comput. Biol. Drug Des. **2**(2), 134–148 (2009). https://doi.org/10.1504/IJCBDD.2009.028825

VerifyThis 2023: An International Program Verification Competition

Xavier Denis[1]([✉]) and Stephen F. Siegel[2][ID]

[1] Université Paris-Saclay, INRIA, ENS Paris-Saclay,
Laboratoire Méthodes Formelles, 91190 Gif-sur-Yvette, France
`denis.xavier@inf.ethz.ch`
[2] University of Delaware, Newark, DE 19716, USA
`siegel@udel.edu`

Abstract. The 11th VerifyThis program verification competition took place on April 22, 2023, at ETAPS 2023 in Paris, France. Contestants were tasked with solving three verification challenges in real time, using any tools of their choice. The subjects of the challenges were (1) in-place reversal of a linked list, (2) ordered binary decision diagrams, and (3) a concurrent FIFO queue. Fifteen teams, each with 1 or 2 members, participated; all were required to be present on site. In this report, we summarize the three challenges and some of the notable solutions.

1 Introduction

VerifyThis is a series of program verification competitions. It has taken place annually since 2011, except for 2013 and 2020. Since 2015, it has been held as a workshop at the European Joint Conferences on Theory and Practice of Software (ETAPS). The 2023 event was the 11th in the series. It took place on Saturday, April 22, in the library of the Institut Henri Poincaré, Sorbonne University, Paris. Sunday was used to present and discuss solutions, and for judging.

In contrast with several other competitions, VerifyThis is not a fully automated affair. Instead, participants are required to be physically present, working in teams of one or two members each. They are given a sequence of three *challenges* to solve in real time. The typical challenge starts with some combination of: a natural language description of a problem, pseudocode, and/or actual code. This is followed by a sequence of verification *tasks*. The tasks usually involve (re-)implementing the algorithm in a language appropriate for the verification tool(s) the participant will use, and then formalizing and verifying certain properties of the implementation. Teams may use any languages and verification tool or tools they wish. Each challenge lasts 90 min. (Some extra time was allowed for Challenge 2 this year).

The competition is therefore measuring the skill of the participants as well as the effectiveness of the tools. As the solutions may use tools with very different input languages, techniques, and kinds of correctness guarantees, there is no straightforward algorithm to rank the solutions. Instead, judges evaluate the solutions based on correctness, completeness, and elegance.

D. Beyer et al. (Eds.): TOOLympics 2024, LNCS 14550, pp. 147–159, 2025.
https://doi.org/10.1007/978-3-031-67695-6_5

```
struct Cell { int value; struct Cell *next; };
struct Cell *list_reversal(struct Cell *l) {
  struct Cell *r = NULL;
  while (l != NULL) {
    struct Cell *tmp = l;
    l = l->next;
    tmp->next = r;
    r = tmp;
  }
  return r;
}
```

Fig. 1. In-place reversal of linked list.

This year, 23 people participated in 15 teams. Participants represented institutions from at least seven countries: Canada, France, Germany, the Netherlands, Switzerland, the United States, and the United Kingdom. They used a variety of tools, virtually all of them based on deductive verification approaches. The tools include Why3, TrustInSoft Analyzer, VerCors, Gobra, Viper, SPARK, SecC, Creusot, and GhostPtrToken. Some of these tools also used automated theorem provers, including Z3, Alt-ergo, and CVC4. Viper appears to have been the most popular tool, with at least 5 teams using it on at least one challenge.

The organizers, who are the authors of this report, solicited challenges from the community and composed some of their own. In the end, we selected one contributed challenge, Challenge 1, proposed by Jean-Christophe Filliâtre and Andrei Paskevich. Challenges 2 and 3 were written by us.

In the remainder of this report, we summarize the challenges and highlight some of the approaches and solutions. Conclusions and a list of the winners are given in the final section.

2   Challenge 1: In-place Reversal of Linked List

The first problem, contributed by Jean-Christophe Filliâtre and Andrei Paskevich, dealt with a standard linked list algorithm: in-place reversal of a singly-linked list. The challenge provided C code implementing this algorithm, shown in Fig. 1.

The algorithm clearly terminates on a NULL-terminated list, but less well-known is the fact that it also terminates on any infinite list that loops at some point, i.e., a lasso-shaped list with an initial segment (of any nonnegative length) and then a cycle (of any positive length). When applied to a list like this, list reversal reverses the cycle and leaves the initial segment unchanged (by reversing it twice). If memory is finite, any list is either NULL-terminated or loops at some point, and thus list reversal always terminates.

The challenge specified five tasks:

1. Implement list reversal in a language of your choice.
2. Show that your implementation, if invoked on a well-formed list, is free of runtime violations; in particular, no invalid pointer operation occurs. Part of this task is to specify precisely "well-formed." (Both NULL-terminated and lasso-shaped lists are considered well-formed.)
3. Show that your implementation, if invoked on any well-formed list in finite memory, terminates.
4. Show that, at termination, the resulting list is the reverse of the original. For lasso-shaped lists, this means the initial segment is unchanged and the cycle is reversed, as explained above.
5. Show that the space and time consumed are linear in the length of the list.

Comments on Solutions. Despite its similarity to the classic introductory separation logic problem of reversing a linked-list, the introduction of a lasso at the end of the list makes this problem much harder to verify. Most teams (11 of 16) verified memory safety, fewer teams proved termination (5 of 16) though 2 additional teams demonstrated portions of the termination argument. One team each proved problems 3 and 4, while a single team proved the linear space usage of the algorithm.

Team *Wondering How and Why3* (Josué Moreau and Paul Patault) verified tasks (1), (2), and (3). An excerpt of their solution is shown in Appendix A. This solution is notable both for the amount of problems solved, but also for the tool used to do it. Why3 does not support separation logic, instead the team manually encoded a Burstall-Bornat memory model [2] to represent their heap. They then specified their lists using two representation predicates. The first predicate, segList2L(mem, s, i, j, l, p), takes a memory mem, an *abstract value* for the list mapping indices of the list to their memory location, a start index i, a final index j, a starting location l, and a loop start index p. The predicate establishes that $s[i..j]$ is a list starting at l and either ending at a null (and thus lasso-free list) or at the position $s[p]$ of the lasso head. The second predicate segList2R(mem, s, i, j, l) takes a memory, an abstract list s, a start index i, final index j and end location l. It establishes that $s[i..j]$ forms a list in memory, but starting from the final cell of the list l. Their loop starts with the whole list represented by segList2R but at each iteration removes one element from that predicate which is then captured by segList2L.

A solution by the authors of this challenge can be found at https://toccata.gitlabpages.inria.fr/toccata/gallery/linked_list_rev.en.html.

3 Challenge 2: Binary Decision Diagrams

The second challenge consisted of verifying a minimal Reduced Ordered Binary Decision Diagram (ROBDD) [1,3] library. BDDs allow compact representations of propositional formulas by sharing common sub-expressions. Formulas are stored as directed acyclic graphs, where each internal node represents the conditional test of a variable x. Two leaf nodes are used to indicate the values $\top$ and $\bot$. Using a BDD, the formula $(x_1 \wedge x_2) \vee x_3$ would be represented as the following graph:

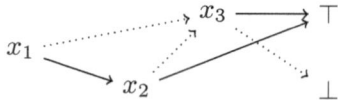

Dotted lines correspond to the false branch of a node, while solid ones correspond to the true branch.

A BDD is *ordered* if all its variables appear in the same order along each path. A BDD is *reduced* if neither of the following transformations can be applied:

1. Eliminate a node with isomorphic children
2. Combine isomorphic subgraphs.

Starter code was provided in three languages: C, Java, and OCaml, the OCaml version is included in Appendix B. The challenge focuses on reasoning about memory aliasing and abstraction through the following tasks:

1. Verify the memory safety of all operations
2. Verify the termination of all operations
3. Verify the operations are correct:
   (a) Applying mk_and(a, b) should produce a node equivalent to the conjunction of the formulas represented by a and b
   (b) Applying not(a) should produce a node equivalent to the negation of the formulas represented by a
4. (a) Verify that each operation preserves the properties of a ROBDD
   (b) If the input BDD is ordered, then the output of each operation is an ordered BDD
   (c) Similarly, if the input BDD is reduced, then the output is also reduced.

Comments on Solutions. This challenge proved to be too long for the time frame set out in VerifyThis. Though several teams were able to prove properties (1) and (2), they ran out of time to verify the other properties. Proving those properties required establishing that the formula memoized by the BDD is equivalent to the original input (i.e., that A is equivalent to $A \land A$). Depending on the implementation of memoization this property could prove challenging to establish.

Team *Morpho Labs* (Quentin Garchery) demonstrated property (3) using Why3. Their solution uses a verified implementation of hashmaps provided by Why3. Usage of this preexisting library allowed them to progress more rapidly. To verify (3), they provided an interpretation function interp(f, n) -> bool which interprets the BDD rooted at node n using a context f mapping variables to true values. This allows them to abstract over the concrete representation of the BDD to focus on its contents. For example, the specification for mk_not is

```
let rec mk_not (b : bdd) (n : node) : node
    variant { n }
    ensures { forall f. interp f result = not interp f n }
```

which states that the BDD found at the node result contains the negation of the one at n.

4 Challenge 3: Nonblocking Concurrent Queue Using LL/SC Synchronization

The third challenge dealt with a concurrent FIFO queue, due to Claude Evéquoz, specified using the load-linked/store-conditional (LL/SC) synchronization instructions [4]. As explained in [4], a variable v accessed by these instructions "... can be regarded as a variable that has an associated shared set of thread identifiers" valid_v, which is initially empty. Each of the two instructions behaves like a function call which possibly modifies the valid set and returns a value, in a single atomic step. The semantics are specified in pseudocode as follows:

$$\mathsf{LL}(v) \equiv \begin{array}{l} \mathsf{valid}_v \leftarrow \mathsf{valid}_v \cup \{tid\}; \\ \textbf{return } v; \end{array} \qquad \mathsf{SC}(v, x) \equiv \begin{array}{l} \textbf{if } tid \in \mathsf{valid}_v \textbf{ then} \\ \quad \left| \begin{array}{l} \mathsf{valid}_v \leftarrow \varnothing; \\ v \leftarrow x; \\ \textbf{return } true; \end{array} \right. \\ \textbf{else return } false \end{array}$$

where tid is the ID of the thread executing the instruction. Specifically, evaluation of $\mathsf{LL}(v)$ results in the value stored in v, but also has the side-effect of adding tid to the valid set associated to v. The result of evaluating $\mathsf{SC}(v, x)$ depends on whether tid was in the valid set of v: if so, the valid set is cleared, the value of x is stored in v, and the evaluation result is $true$. If tid was not in the valid set of v, then the evaluation result is $false$ and no change is made to v or its valid set.

The queue is implemented using an array Q as a cyclic bounded buffer. C code for the *enqueue* operation is shown in Fig. 2. The code assumes the element type is int and that only nonnegative integers are added to the queue. The integer -1 is used to represent a "null" value. The challenge also noted the following:

– Initially, Head = Tail = 0 and $Q[i]$ = null = -1 for $0 \le i <$ LEN.
– Head and Tail increase monotonically; for this challenge, assume the unsigned int type is unbounded.
– The number of elements stored in the queue is Tail − Head, and these elements are located at positions Head%LEN, (Head + 1)%LEN, ..., (Tail − 1)%LEN of Q.
– Assume a sequentially consistent memory model, i.e., an execution is an interleaved sequence of atomic actions from the different threads, and the value read from a memory location is the last value written to that location.

The challenge posed the following tasks:

1. In your favorite language, write a program P incorporating Evéquoz's FIFO queue (only the enqueue operation is needed). The queue is initially empty. P generates NT threads (NT ≥ 1), with IDs $0, \ldots,$ NT $- 1$. Each thread calls *enqueue* on its thread ID, then terminates.
2. Show that all executions of P terminate (i.e., all threads terminate).
3. Show that no out-of-bound array indexes occur on any execution of P.
4. Assuming NT $\le$ LEN, show that in any execution of P,

```
int Q[LEN];  // array with indexes in 0..LEN-1
unsigned int Head, Tail;
bool enqueue(int val) {
  unsigned int t, tailSlot;
  int slot;
  while (true) {
    t = Tail;
    if (t == Head + LEN) return false; // queue is full
    tailSlot = t % LEN;
    slot = LL(&Q[tailSlot]);
    if (t == Tail) {
      if (slot != null) {
        if (LL(&Tail) == t) SC(&Tail, t+1);
      } else if (SC(&Q[tailSlot], val)) {
        if (LL(&Tail) == t) SC(&Tail, t+1);
        return true; // success
      }
    }
  }
}
```

Fig. 2. Nonblocking concurrent queue using LL/SC: enqueue operation.

   (a) all calls to **enqueue** return *true* (success);
   (b) at the final state, the size of the queue (Tail − Head) is NT;
   (c) at the final state, the contents of the queue are some permutation of the integers $0, \ldots, NT - 1$.
5. Assuming NT > LEN, show that in any execution of P,
   (a) at the final state, the queue is full (Tail − Head = LEN)
   (b) the data in the queue is some permutation of a subset of size LEN of $0, \ldots, NT - 1$.

(A second part, dealing with the *dequeue* function, was provided in case anyone finished the first part with time to spare. No one did.)

Comments on Solutions. This proved to be the most difficult of the three challenges. In part, this is due to the technology that participants used. The deductive verification tools have limited ability to reason about concurrency. When developing this problem, we had in mind model checking tools, such as Spin [5] or the CIVL Model Checker [6]. Most such tools use languages in which concurrent algorithms can be expressed naturally. Moreover, the reasoning required to check properties is mostly automatic, albeit limited to a finite (usually small) scope.

    An excerpt from a CIVL solution to this challenge is shown in Appendix D. The language is primarily C, with additional primitives with names that begin with $. The shared objects are represented using a C struct that bundles an integer value with a bit set representing the set of threads in the object's *valid* set. The LL and SC functions are defined in the scope of function `thread`, so that the thread ID `tid` is in scope. Function `enqueue` is almost a verbatim copy of the given code. The model checking engine exhaustively explores all interleavings and checks that the assertions (not shown) never fail. For this model, task 4 can be verified for NT = LEN = 4 in about 25 s. Task 5 verifies in under 10 s for NT = 3 and LEN = 4.

Several teams were able to complete or nearly complete tasks 1 and 3. One team was able to show termination for a single thread execution, but in this case the loop terminates after only one iteration.

A classic approach to proving properties of concurrent programs is to show some property of the state is invariant under every atomic action. One team, *Somehow Alex & Marco Returned* (Alex Summers and Marco Eilers), made an interesting attempt along those lines, excerpted in Appendix E. The concurrent program is essentially modeled as a sequential program with a nondeterministic scheduler. The state of a thread is represented using the original variables (`Head`, `Tail`, etc.) together with an integer *program counter* variable `pc`, which can take on approximately 12 values corresponding to specific locations in the code. To execute one atomic step, a thread is chosen nondeterministically. The program switches on the value of the thread's `pc` to a case that performs a single atomic update to the state. This is repeated until no thread is enabled. The goal was to verify that a number of global invariants were preserved at each iteration of this loop. While the solution was not complete, the approach is interesting and we see no reason it could not be extended to a full solution. Of course, it required manual effort to translate the given code to an explicit state machine, and this translation could also introduce errors. If this process could be automated, it could yield an effective approach to deductive verification of concurrent programs.

5 Closing Remarks

Like previous years, the competition awarded several prizes to reward participants. This year the prizes were:

- **Best overall team**: *Somehow Alex & Marco Returned*, Alexander J. Summers and Marco Eilers
- **Best contributed problem**: Jean-Christophe Filliâtre and Andrei Paskevich
- **Best student team**: *Loops and Dots*, Jonas Fiala and Thibault Dardinier
- **Most interesting tool feature**: *Wake Me Up When Verification Ends*, Linard Arquint and Joao Pereira, for the "first-class predicates" of Gobra.

The competition showcased the power of current verification technology, as well as the creativity of the participants. However, it also revealed areas where improvement is needed. Reasoning about pointers and memory continue to challenge the deductive verification tools commonly used at VerifyThis. While linked-list based problems are standard fare in imperative program verification, an apparently small twist—allowing the list to terminate in a cycle—required predicates, variants and invariants that were very difficult to formulate, especially in separation logic. Reasoning about the memory sharing of BDDs was found to be even more challenging.

Concurrency also remains a stumbling block. The deductive verification tools used at VerifyThis need better ways to express and reason about concurrent programs. Techniques that allow one to easily express a global invariant, and prove that it remains invariant under each atomic action taken by a thread, would be one approach. It may also be the case that students and practitioners of program verification need more practice verifying concurrent algorithms.

In contrast, finite-state verification techniques, such as model checking, are particularly effective at reasoning about concurrent systems, and are generally easily automated. Of course, the downside is that such techniques typically only "prove" that properties hold in some finite scope. The two approaches—deductive and finite-state—have complementary strengths, and a team with expertise in both would certainly do very well at VerifyThis.

As in prior years, the participants of VerifyThis 2023, and other members of the community, have been invited to submit revised or new solutions to the challenges, which will be posted on the Archive of the VerifyThis web site [7].

Acknowledgements. Financial support for the prizes was provided by Amazon Web Services. S.F. Siegel was supported by the U.S. National Science Foundation under awards CCF-1955852 and CCF-2019309.

A   Challenge 1 Solution

The following is an excerpt from solution to Challenge 1 by team *Wondering How and Why3* (J. Moreau and P. Patault):

```
let rev (ref l: loc) (ghost s: int -> loc) (ghost len p: int): loc
  requires { valid s len }
  requires { len >= 0 }
  requires { segList2R mem s 0 len l p }
  ensures  {
    (old mem).next (s (len-1)) = null -> (* case: non lasso *)
      segList2L mem s 0 len result }
  ensures  {
    (old mem).next (s (len-1)) <> null -> (* case: lasso *)
      segList2R mem s 0 p l p /\ (* initial order until loop *)
      segList2L mem s p len l }
=
  let ref r = null in
  let ghost ref i = 0 in
  while l <> null do
    invariant { segList2R mem s i len l p }
    invariant { segList2L mem s 0 i r }
    invariant { i <= len }
```

```
  variant   { len - i }
  let ref tmp = l in
  l <- get l;
  set tmp r;
  r <- tmp;
  i <- i + 1;
done;
r
```

B   OCaml Starter Code for Challenge 2

```
type var = int
type node = Node of { var: int; left: node; right: node; } | True | False

let equal n m = match n, m with
  | True, True -> true
  | False, False -> true
  | Node n, Node m -> n.var = m.var && n.right == m.right &&
                      n.left == m.left
  | _ -> false

module Tbl = Hashtbl.Make(struct
  type t = node
  let equal = equal
  let hash = Hashtbl.hash
end)
type bdd = node Tbl.t
let mk_bdd () : bdd = Tbl.create 32

let mk_node (b : bdd) (n : node) : node = match Tbl.find_opt b n with
  | Some n -> n
  | None -> Tbl.add b n n; n

let mk_true b : node = mk_node b True
let mk_false b : node = mk_node b False

let mk_if b var left right =
  if equal left right then left else
  mk_node b (Node { var; left; right })

let mk_var b v = mk_if b v (mk_true b) (mk_false b)

let rec mk_not (b : bdd) (n : node) : node = match n with
  | True -> mk_false b
  | False -> mk_true b
  | Node { var; left; right } ->
      mk_if b var (mk_not b left) (mk_not b right)

let rec mk_and (b : bdd) (l : node) (r : node) : node = match l, r with
```

```
| True, _ -> r
| _, True -> l
| False, _ | _, False -> mk_false b
| Node {var = vara; left = lefta; right = righta },
  Node { var = varb; left = leftb; right = rightb } ->
    begin match compare vara varb  with
    | -1 -> mk_if b vara (mk_and b lefta r) (mk_and b righta r)
    | 0 -> mk_if b vara (mk_and b lefta leftb) (mk_and b righta rightb)
    | 1 -> mk_if b varb (mk_and b l leftb) (mk_and b l rightb)
    | _ -> assert false
    end
```

C   Challenge 2 Solution

The following is extracted from team *Morpho Labs*'s (Q. Garchery) submission
for Challenge 2:

```
type var = int
type node = Node var node node | T | F

let rec ghost function interp (f: var -> bool) node
= match node with
    | T -> true
    | F -> false
    | Node var left right ->
        if f var then interp f left
        else interp f right
    end

val equal (a b : node) : bool ensures { result <-> a = b }

clone hashtbl.Hashtbl as Tbl with type key=node
type bdd = Tbl.t node

let mk_node (b : bdd) (n : node) : node
    ensures { result = n }
= if not Tbl.mem b n then Tbl.add b n n;
    return n

let mk_true b : node
 ensures { forall f. interp f result = true }
= mk_node b T

let mk_false b : node
 ensures { forall f. interp f result = false }
= mk_node b F

let mk_if b var left right
    ensures { forall f. interp f result =
```

```
      if f var then interp f left else interp f right }
= if equal left right then left else
    mk_node b (Node var left right)

let rec mk_not (b : bdd) (n : node) : node
    variant { n }
    ensures { forall f. interp f result = not interp f n }
= match n with
    | T -> mk_false b
    | F -> mk_true b
    | Node var left right -> mk_if b var (mk_not b left) (mk_not b right)
  end
```

D   CIVL Model of Evéquoz queue

The following is an excerpt from the CIVL model of the Evéquoz queue. The main function (not shown) spawns NT threads, each executing thread. When $NT \leq LEN$, after all return, it checks the contents of Q is some permutation of $0..NT - 1$. The complete solution is available at [7], in the 2023 Archive.

```
typedef struct fint { // an int bundled with a valid set
  int data;
  _Bool valid[NT];
} fint;
fint Q[LEN], Head, Tail; // shared variables
void thread(int tid) {
  $atomic_f int LL(fint * x) { // linked-load operation
    x->valid[tid] = 1;
    return x->data;
  }
  $atomic_f _Bool SC(fint * x, int val) {
    // store conditional operation
    if (x->valid[tid]) {
      for (int i=0; i<NT; i++) x->valid[i]=0;
      x->data = val;
      return 1;
    } else { return 0; }
  }
  _Bool enqueue(int val) { // attempts to enqueue val
    int t, tailSlot, slot;
    while (1) {
      t = Tail.data;
      if (t == Head.data + LEN) return FULL_QUEUE;
      tailSlot = t % LEN;
      slot = LL(&Q[tailSlot]);
      if (t == Tail.data) {
        if (slot != null) {
          if (LL(&Tail) == t) SC(&Tail, t+1);
        } else if (SC(&Q[tailSlot], val)) {
```

```
            if (LL(&Tail) == t) SC(&Tail, t+1);
            return OK;
          }
        }
      }
    }
  enqueue(tid); // Body of thread function
}
```

E   Challenge 3: Explicit Sequentialization

The following excerpt is from the solution to Challenge 3 submitted by A. Summers and M. Eilers. It shows the use of Viper's statement macros to define LL and SC, and the explicit sequentialization of the concurrent enqueue function.

```
field valid: Set[Int]
...
define LL(tid, v, target) {
  v.valid := v.valid union Set(tid)
  target := v.val
}
define SC(tid, v, x, target) {
  if (tid in v.valid) {
    v.valid := Set()
    v.val := x
    target := true
  } else {
    target := false
  }
}
...
  while ... {
    var tid : Int
    tid := chooseThreadID(); // pick
    var T : Ref := threads[tid]
    var PC : Int := T.pc
    if(PC == 1) {
      T.t := Tail.val
      T.pc := 2
    } elseif (PC == 2) {
      if(T.t == Head.val + LEN()) {
        T.pc := 11
      } else {
        T.tailSlot := T.t % LEN()
        T.pc := 3
      }
    } elseif (PC == 3) {
        var ll : Int
        LL(tid,loc(Q,T.tailSlot),ll) // LL(&Q[tailslot])
```

```
        T.slot := 11
        T.pc := 4
    }
    ...
}
```

References

1. Akers: Binary decision diagrams. IEEE Trans. Comput. **C-27**(6), 509–516 (1978). https://doi.org/10.1109/TC.1978.1675141
2. Bornat, R.: Proving pointer programs in Hoare logic. In: Backhouse, R., Oliveira, J.N. (eds.) MPC 2000. LNCS, vol. 1837, pp. 102–126. Springer, Heidelberg (2000). https://doi.org/10.1007/10722010_8
3. Bryant, R.E.: Graph-based algorithms for boolean function manipulation. IEEE Trans. Comput. **100**(8), 677–691 (1986). https://doi.org/10.1109/TC.1986.1676819
4. Evéquoz, C.: Practical, fast and simple concurrent FIFO queues using single word synchronization primitives. In: Kordon, F., Vardanega, T. (eds.) Ada-Europe 2008. LNCS, vol. 5026, pp. 59–72. Springer, Heidelberg (2008). https://doi.org/10.1007/978-3-540-68624-8_5
5. Holzmann, G.J.: The Spin Model Checker. Addison-Wesley, Boston (2004)
6. Siegel, S.F., et al.: CIVL: the concurrency intermediate verification language. In: SC'15: Proceedings of the International Conference for High Performance Computing, Networking, Storage and Analysis, article no. 61, pp. 1–12. ACM, New York (2015). https://doi.org/10.1145/2807591.2807635
7. VerifyThis Competition (web site). http://verifythis.ethz.ch. Accessed 14 Feb 2024

The VerifyThis Collaborative Long-Term Challenge Series

Wolfgang Ahrendt[1], Gidon Ernst[2(✉)], Paula Herber[3], Marieke Huisman[4],
Raúl E. Monti[4], Mattias Ulbrich[5], and Alexander Weigl[5]

[1] Chalmers University of Technology, Gothenburg, SE, Sweden
ahrendt@chalmers.se
[2] LMU Munich, Munich, Germany
gidon.ernst@lmu.de
[3] University of Münster, Münster, Germany
paula.herber@uni-muenster.de
[4] University of Twente, Enschede, The Netherlands
{m.huisman,r.e.monti}@utwente.nl
[5] Karlsruhe Institute of Technology, Karlsruhe, Germany
{ulbrich,weigl}@kit.edu

Abstract. We give a brief overview of the VerifyThis long-term challenge series. Goal of these challenges is to demonstrate practical value of formal methods, to evaluate the current tools on specifying and verifying requirements of realistic software systems, and to bring together the community for an exchange on the state-of-the-art and future directions. An emphasis is placed on encouraging collaboration between participating research groups, not just at a conceptual level but also towards integrating verification tools and approaches, e.g., sharing technical artifacts such as specifications and proofs.

Website: https://verifythis.github.io/
Mailing List: verifythis-ltc@lists.kit.edu

Keywords: Software Systems · Specification · Verification

1 Motivation

The VerifyThis long-term challenges are a community-driven effort in the area of formal methods for software systems. The challenges comprise the specification and verification of various properties of interest for real-world open-source software components. The motivation for this effort comes out of the community around the VerifyThis [8] and VSComp [18] competition events. In these competitions, small but intricate verification problems are solved competitively by participating teams within a short time (90 min up to 2 days, typically on-site), using a tool of their choice. Hereby, human guidance is essential for writing specifications and for providing proof hints such as invariants and data abstractions [8,15]. This aspect, which is also important in the long-time challenges,

D. Beyer et al. (Eds.): TOOLympics 2024, LNCS 14550, pp. 160–170, 2025.
https://doi.org/10.1007/978-3-031-67695-6_6

contrasts most of the other competitions at TOOLympics which instead evaluate the performance of fully automatic tools.

While fun and insightful, to be tractable under the conditions of the short competitive VerifyThis events, the problems considered there are limited in scope and typically consider data structures and algorithms in isolation only. Therefore this setting is not suitable for a more realistic evaluation of methods and tools on larger code bases and full systems, which motivated this series originally [16]:

> *What could be achieved in the area of program verification if*
> *(a) we as the program verification community collaborated and*
> *(b) the time constraints were removed?*

The long-term challenges are thus similar in spirit to the Mondex challenge [24] and NASA's flash file system challenge [17], which had been put forward in response to Tony Hoare's grand vision of a verifying compiler [14].

This paper summarizes scope and goals (Sect. 2) as well as the efforts around two past challenges, related to the Hagrid keyserver [16] (Sect. 3.1) and a Casino smart contract (Sect. 3.2). We call for participation for the current challenge, the verification of the key-value store memcached [10] (Sect. 3.3), before we summarize the outcomes and insights (Sect. 4).

2   Scope, Goals, and Organization

VerifyThis long-term challenges are intended to involve researchers from different but overlapping communities and as such are set up broadly in terms of the type of requirements being formalized and the specification and verification techniques used (e.g. automata, contracts, temporal logic, refinement, model checking, deductive verification, static analysis, testing, . . .).

Goals of the long-term challenges include

1. to foster collaboration between researchers and their tools,
2. to demonstrate practical value of formal methods, and
3. to evaluate and improve the capabilities and maturity of methods and tools.

The emphasis on collaboration comes with the need and at the same time the opportunity to make progress on long-standing open issues in formal methods [13]

4. to develop approaches that bridge between specification paradigms and
5. to work towards conceptual and technical integration of verification tools.

Considerable thought goes into the selection and preparation of the challenge subjects. Challenges are proposed by members of the community, suggesting key concerns to look into as well as preparing the necessary background information for others to work on the challenge. Concrete **criteria for good long-term challenges** have been formulated alongside the first challenge proposal in [16, Sec 3.1], which ensure that the planned efforts are in scope and that the goals outlined are achievable in a meaningful way. We summarize these briefly:

- the challenge should be based on a real-world, open source software
- there should be a wide range of correctness concerns that span from the global systems level perspective down to properties of the source code, and
- the challenge is modular with some well-understood core functionality or component and further more complex requirements and parts

Due to these characteristics, challenge problems tend to be much more open-ended in comparison to the benchmarks and concrete verification tasks of other TOOLympics events. In particular, defining the scope of activities and formulating the respective requirements is up to the participants. Similarly, while verifying the existing implementation is a desirable target, a more viable alternative that is frequently chosen by participants is to implement a system with (a subset of) the desired functionality from scratch. This allows them to focus on aspects deemed interesting from a research perspective. At the same time this flexibility leads to a larger variety of solutions so that a *competitive* evaluation with respect to pre-established metrics is not a goal of this series (it is already difficult for the VerifyThis on-site events [8]).

3   Challenges and Solutions

We briefly summarize the proposed challenges and solutions, insights and conclusions are drawn in summary in Sect. 4. The first challenge was proposed online, solutions were presented at a virtual workshop [21]. The second challenge was mostly discussed in a series of online-events, which were open to and attended by people interested but not directly involved in working on the proposed problems. The discussions at these meetings are summarized at https://verifythis.github.io/events/. For the current challenge, dedicated in-person meetings are planned (Sect. 3.3). The time-frame for each challenge is roughly one to two years, but as it is the case with the shorter VerifyThis events, some groups continue to work on them or follow-up on research ideas that came out the discussions.

3.1   Verifying Key-Server `Hagrid` (2019–2020)

Challenge Proposal: M. Huisman, R. Monti, M. Ulbrich, and A. Weigl [16]
Challenge Website: https://verifythis.github.io/01hagrid/
Reference System: https://keys.openpgp.org/
                         https://gitlab.com/hagrid-keyserver/

Description. The subject of the first challenge was the "verifying" key-server `Hagrid`. Its purpose is to store a registry of public PGP keys used for encryption of E-Mails. Such keys can be uploaded by users and optionally *validated* to be discoverable via associated E-Mail addresses. In contrast to older similar services, `Hagrid` implements a validation step via a confirmation e-mail that ensures that such associations are genuine, thereby protecting against maliciously placed entries. The schematics of `Hagrid` are shown in Fig. 1.

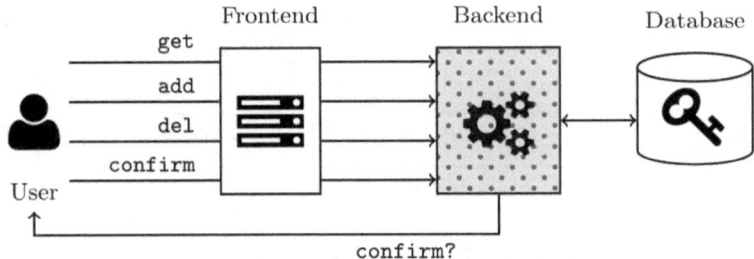

Fig. 1. Schematic of the key server architecture [16, Fig. 1], where the front-end manages the protocol-side of user requests (e.g., https://keys.openpgp.org/ offers a web interface and a JSON API), the back-end implements the functional behavior (its specification comprises the "core" of the challenge), and the database implements key storage (in Hagrid it is realized on the file system).

The challenge fits well with the goals set out in Sect. 2, as it encompasses at its core a specification of its functional behavior that can be extended towards different requirements (e.g. security aspects such as proper use of randomness, privacy and non-leakage, authenticity of entries, or performance characteristics) that can be connected towards an implementation (e.g., of the storage back-end, of the communication protocol). Possible requirements to be specified and verified were suggested in the challenge proposal as so-called "missions", each which represents a desirable property of Hagrid.

Contributions. Solutions to the challenge were presented and discussed in an online workshop, the format was chosen due to the pandemic situation preventing travel. There were five contributions submitted to the informal proceedings [21], as summarized in [16, Sec 5.1]:

- Claire Dross, Johannes Kanig and Yannick Moy: A Solution to the Long-Term Challenge in SPARK
- Diego Diverio, Cláudio Lourenço, and Claude Marché: "You-Know-Why": an Early-Stage Prototype of a Key Server Developed using Why3
- Stijn de Gouw, Mattias Ulbrich, and Alexander Weigl: The KeY Approach on Hagrid
- Gidon Ernst and Lukas Rieger: Information Flow Testing of a PGP Keyserver
- Gidon Ernst, Toby Murray, and Mukesh Tiwari: Verifying the Security of a PGP Keyserver

Several contributions established functional correctness with contracts of the top-level operations over a logical model of the internal state ("ghost code"), e.g. by the SPARK, Why3, and KeY teams. The verification of these contracts encompasses safe execution and usually also termination (missions 1, 2 resp. 4, and 8 in [16]). Proper use of randomness was not looked into (mission 6) and none of the solutions is concurrent (mission 7). Two solutions looked into security/privacy issues specifically (mission 5), one based on testing and one based

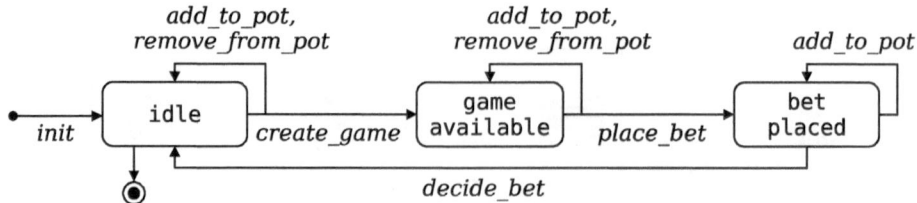

Fig. 2. High-level state machine transitions of the game [9, Fig. 5].

on deductive proofs in SecC. The Why3 solution comes with a verified executable server but it implements a much simpler protocol than that of `Hagrid` (mission 4).

3.2   Casino Smart Contract (2021–2022)

Challenge Proposal: W. Ahrendt
Challenge Website: https://verifythis.github.io/02casino/

Overall, 9 contributions to the Casino challenge will be documented in a joint article that is currently in preparation to be submitted to the theme "Competitions and Challenges" of the Journal "Software Tools for Technology Transfer" (STTT).

Description. The Casino smart contract challenge considers a program running on the Ethereum blockchain, which implements a guessing game. In this particular game, an operator offers the opportunity to players to bet on the outcome of a coin toss in virtual currency. Verifying smart contracts incurs additional challenges that come with the execution model of the blockchain in comparison to typical software: Not only the code is public but so is its data during execution, the participants in the game may have confidence in the source code of the Casino contract but they would not necessarily trust each other. A key component of the challenge is therefore to ensure that the game plays out fairly regardless of any malicious action by either of them, which is achieved in a correct implementation by adhering to a protocol guarding the pot (to ensure that prizes can be paid out, Fig. 2) and to a cryptographic protocol that enforces that the winning coin toss is fixed upfront but remains secret until after the bet has been placed and the outcome of the is revealed.

The source code of the Casino contract provided for the challenge included some known bugs, representative to typical issues found in real blockchain code, e.g., due to possible re-entrancy, information leakage, and possibilities in foreign code to disrupt successful completion of own operations. In contrast to the Hagrid and Memchached challenges, the Casino challenge was thereby not based on a larger existing reference system, but rather focused on these specific and somewhat atypical aspects of verification.

Contributions. Around 10 solutions and approaches have been presented at the informal online events, some of which are linked on the challenge website, including contributions using UPPAAL, TLA⁺, SolC-Verify, SecC, VerCors, JavaBIP, 2vyper, which will be documented an upcoming joint article. Overall, this challenge turned out to be more open to variations in the type of properties considered as well as in the methods used for the analysis, for a number of reasons: First, the execution semantics of smart contracts is different from traditional programs, as such necessitating additional means for specification and verification (cf. challenge description). Second, the group of participants was more heterogeneous than that for the key-server challenge, which brought in diverse viewpoints and therefore positively contributed to the discussions and inspired follow-up activities.

The contributions encompass high-level models, e.g., in UPPAAL, TLA+ or Event-B, which are more detailed versions of the state machine in Fig. 2, and approaches that connect specifications to implementation code (e.g. in C, Java, and the PVL language of VerCors). Other solutions are realized in tools that directly support verification of smart contracts, namely Supremica, SolC-verify, and 2Vyper.

3.3 Key-Value Store memcached (2023–)

Challenge Proposal:	G. Ernst and A. Weigl [10, 11]
Challenge Website:	https://verifythis.github.io/03memcached/
Reference System:	https://memcached.org/
	https://github.com/memcached/memcached

Preliminary executable models of partial functional behavior of memcached are available for Python[1] and for Java[2].

Description. memcached is a key-value storage that acts as a data cache and that is deployed as part of web and cloud software, among its users are many major companies and websites. The system is realized as a client-server architecture, as shown in Fig. 3. The server-side implements the cache storage and entry lifecycle, entries are deleted either upon a user-specified expiry timeout or under memory pressure. Client-side libraries implement programming language-specific APIs over the text-based network interface offered by the server.

Like the first challenge on Hagrid (Sect. 3.1), formally specifying and verifying memcached encompasses different aspects and types of requirements. Its core functionality admits a concise functional description, but there are high-level properties suitable for abstract models and temporal analysis, and the reference implementation comes with low-level concerns like memory management, pointer-structures, and concurrency.

---

[1] https://github.com/gernst/pycached.
[2] https://github.com/wadoon/bloatcache.

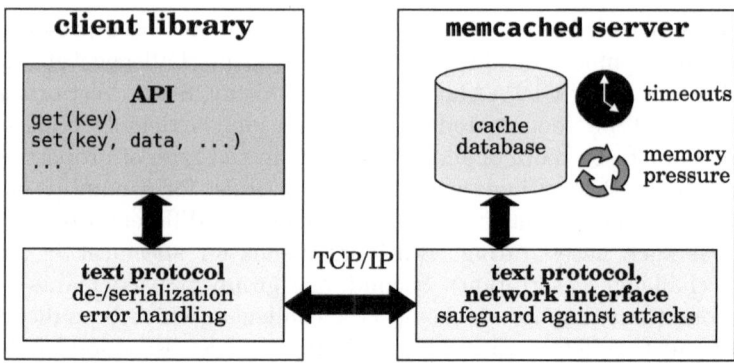

Fig. 3. Schematic of the client-server architecture of memcached [10, Fig. 1].

Call for Participation. We call for contributions of any kind to this challenge. Participants are kindly requested to register via a mailing list (see abstract) and to actively seek out collaboration on specific topics. A first informal in-person meeting will take place at iFM 2023 in Leiden, a follow-up session is planned to be integrated into the VerifyThis workshop at ETAPS 2024. We intend to again host the "SpecifyThis" track at ISoLA 2024 with formal proceedings that specifically welcomes contributions to this challenge.

4   Discussion and Outlook

We discuss the success of the VerifyThis long-term collaborative challenge series with respect to the goals outlined in Sect. 2.

Foster Collaboration Between Researchers. As a community-building measure, the VerifyThis long-term series was very successful. The online meetings were consistently attracting around 20–40 participants and sparked lively and insightful discussions.

Talking about a concrete challenges turned out to be essential as a basis for a common understanding of the problem being solved as well as the respective properties of interest. It also paved the way for the transition from the first to second challenge, during which principles of specification languages were discussed (goal 4, see also the discussion on bridging paradigms) and which prompted the selection of the Casino smart contract as the subsequent subject.

Exchange of ideas occurred at a conceptual level, e.g., the top-level specifications for Hagrid are all very similar, being based on an original model in Event-B. Closer collaboration among participants and a technical integration of solutions was discussed multiple times but ultimately not really attempted. Perhaps a take-away is that this aspect, too, needs more concrete stimulation [16, Sec 5.3], such as maintaining a shared repository for artifacts on a code-hosting platform. We hope that physical events will also help here.

Demonstrate Practical Value of Formal Methods. In comparison to the verification problems solved at VerifyThis competitions, the long-term challenges make a strong effort towards more realistic application of formal methods. A key difference is that the latter consider entire software systems in contrast to isolated data structures and algorithms.

Overall, the solutions are hands-on, have produced actual running code in real programming languages, and deal with the respective application- and domain-specific correctness concerns. However, the resulting verified software for the `Hagrid` challenge is still far from being on-par with the reference system, lacking support for much of the PGP-related concepts. For that reason, the third challenge has been based on a system with a significantly simpler top-level interface, so that it is actually realistic to implement and verify a drop-in replacement for the reference system that is compatible at the protocol interface on the main features of `memcached`. Verifying the actual `memcached` implementation can be considered a realistic albeit ambitious goal given current technology.

Evaluate the Capabilities of Methods and Tools. Functional specifications of behavioral correctness are well-understood and well-supported by deductive verification tools. While they provide powerful means for mathematically precise and scalable reasoning, all existing tools heavily rely on user-provided proof hints such as invariants. Approaches based on automata and temporal logic, on the other hand, are often supported by fully automatic tools, but suffer from the state-space explosion problem. This often forces us to simplify the models to enable verification, as we have seen in the Casino challenge.

Properties that go beyond functional correctness, such as security requirements as well as domain-specific concerns like those that come with blockchain execution semantics, are supported directly by specialized tools only and otherwise need to be modeled or approximated by more traditional means.

Unsurprisingly, there are many aspects that remain challenging. Good proof automation is achievable for fully annotated programs as well as higher-level models, but the automatic inference of loop invariants and coupling relations is currently only done for numeric properties if at all. Making significant progress here appears to require fundamental theoretical and methodological advances. Integration with unverified interfaces, real-world APIs, and external code may be an issue, this is most prominent in the solutions to the `Hagrid` challenge, where formally specifying (and verifying) the actual PGP key format as well as the communication protocol appeared far out of reach of the capacities of the participating teams (cf. the point raised in the discussion of practicality).

Bridging the Gap Between Specification Paradigms. Within the formal methods community, different kinds of specifications have been proposed and applied. For deductive verification, functional specifications are typically written in the form of pre- and postconditions. In the model checking community, specifications are usually defined in temporal logics, and automata are used to describe the system behavior. While pre-/postconditions usually specify the behavior in a static, state-pair fashion and are often better suited to capture program behavior on a detailed level, temporal logics, dynamic logics, and automata also enable the specification of dynamic interaction behaviors and are better-suited for more high-level system properties. While the former kind of specifications was prevalent in the solutions for Hagrid, the latter were widespread in the solutions of the Casino example. We ascribe this to the higher-level implementation of the Casino compared to Hagrid. We could also see that different kinds of properties were addressed with the different specification paradigms. Altogether, this sparked a lot of interesting discussions on the integration of temporal/process-like formalisms into contract-based tools and the other way around. This is an active research topic, and has been much discussed both in the events associated with this series as well as in the literature [1,2,22,23].

Integration of Approaches and Tools. As mentioned, a main obstacle to improving collaboration between participants is the lack of shared artifacts. The main reason is that tools lack the capability to talk to each other, e.g., because they consider different properties and specification paradigms. These differences can be fundamental (e.g. traces vs. contracts, separation logic vs. dynamic frames), or subtle (e.g. semantics of logical data types). Tool compatibility has been a long-standing general issue which is hard to resolve by principle, since development is often driven by exploring *new* ideas that are meant to exceed existing capabilities and thus may require fundamentally new features.

Some reasonably well-defined and widely-used formats for specifications and proofs exist, to name a few: SMT-LIB [4] has been highly influential in the area of automatic theorem provers and its standardization has unlocked a huge potential across a wide range of use-cases. The witness format of SV-COMP [6] can capture some lower-level correctness certificates but lacks many features typically needed to adequately capture functional correctness. JML [20] and ACSL [5] are specification and annotation languages specific for Java resp. C and hence encompass many features to deal with the specifics of these languages. However, what seems to be desirable instead is a format for describing behavioral abstractions of software components independently of the implementation language as well as the verification method and tool.

In between the first and second challenge, two online events were dedicated to this issue of specification paradigms. An actionable observation that has come out of these discussions is that there is ground for a common formalism that captures abstract behavioral models of stateful components, encapsulated objects, or systems with well-defined interfaces in terms of an abstract "ghost state" over which the transitions corresponding to method calls are expressed as part of their contracts. This style of specification is effectively available in most deductive code-level verifiers, and it has the advantage of offering a well-defined point for integration, not just across these tools, but also as a basis for bridging paradigms (cf. discussion of the previous goal). Examples for similar integration efforts are [7] at the level of C assertions and [3] as well as [19] for Java, but there is currently format that is expressive as well as language- and tool-independent.

Outlook. We remark that the long-term challenges and the discussion presented here are not intended as a controlled scientific experiments against objective measures. The problems considered are simply too open-ended and the emphasis on human guidance makes such an evaluation difficult [8]. However, moving to more controlled experiments has been discussed and determined to be a valid goal, but this is currently out of scope and likely requires an appropriate benchmark suite.

The agenda of bridging specification paradigms has been followed up with the "SpecifyThis" track at ISoLA 2022 [1], Dagstuhl seminar 22451 "Principles of Contract Languages" [12], and a Lorentz workshop "Contract Languages". Developing a common interchange format from these ideas, however, is still future work that requires a dedicated effort distinct from this challenge series.

Acknowledgement. We thank the anonymous reviewers for their feedback and suggestions to improve the paper. Huisman is supported by the NWO VICI 639.023.710 Mercedes project.

References

1. Ahrendt, W., Herber, P., Huisman, M., Ulbrich, M.: SpecifyThis - bridging gaps between program specification paradigms. In: Margaria, T., Steffen, B. (eds.) ISoLA 2022. LNCS, vol. 13701, pp. 3–6. Springer, Cham (2022). https://doi.org/10.1007/978-3-031-19849-6_1

2. Amilon, J., Lidström, C., Gurov, D.: Deductive verification based abstraction for software model checking. In: Margaria, T., Steffen, B. (eds.) ISoLA 2022. LNCS, vol. 13701, pp. 7–28. Springer, Cham (2022). https://doi.org/10.1007/978-3-031-19849-6_2

3. Armborst, L., Lathouwers, S., Huisman, M.: Joining forces! Reusing contracts for deductive verifiers through automatic translation. In: Herber, P., Wijs, A. (eds.) iFM 2023. LNCS, vol. 14300, pp. 153–171. Springer, Cham (2023). https://doi.org/10.1007/978-3-031-47705-8_9

4. Barrett, C., Fontaine, P., Tinelli, C.: The SMT-LIB Standard: Version 2.6. Technical report, Department of Computer Science, The University of Iowa (2017). http://smtlib.cs.uiowa.edu/language.shtml

5. Baudin, P., Filliâtre, J.C., Marché, C., Monate, B., Moy, Y., Prevosto, V.: ACSL: ANSI/ISO C Specification Language. http://frama-c.com/download/acsl.pdf

6. Beyer, D.: Competition on software verification and witness validation: SV-COMP 2023. In: Sankaranarayanan, S., Sharygina, N. (eds.) TACAS 2023. LNCS, vol. 13994, pp. 495–522. Springer, Cham (2023). https://doi.org/10.1007/978-3-031-30820-8_29

7. Beyer, D., Spiessl, M., Umbricht, S.: Cooperation between automatic and interactive software verifiers. In: Schlingloff, B.H., Chai, M. (eds.) SEFM 2022. LNCS, vol. 13550, pp. 111–128. Springer, Cham (2022). https://doi.org/10.1007/978-3-031-17108-6_7

8. Ernst, G., Huisman, M., Mostowski, W., Ulbrich, M.: VerifyThis – verification competition with a human factor. In: Beyer, D., Huisman, M., Kordon, F., Steffen, B. (eds.) TACAS 2019. LNCS, vol. 11429, pp. 176–195. Springer, Cham (2019). https://doi.org/10.1007/978-3-030-17502-3_12

9. Ernst, G., Knapp, A., Murray, T.: A Hoare logic with regular behavioral specifications. In: Margaria, T., Steffen, B. (eds.) ISoLA 2022. LNCS, vol. 13701, pp. 45–64. Springer, Cham (2022). https://doi.org/10.1007/978-3-031-19849-6_4

10. Ernst, G., Weigl, A.: Verify This: memcached–a practical long-term challenge for the integration of formal methods. In: Herber, P., Wijs, A. (eds.) iFM 2023. LNCS, vol. 14300. Springer, Cham (2023). https://doi.org/10.1007/978-3-031-47705-8_5

11. Ernst, G., Weigl, A.: VerifyThis Long-term Challenge: Specifying and Verifying a Real-life Remote Key-Value Cache (memcached) (2023). https://verifythis.github.io/03memcached/challenge.pdf

12. Gurov, D., Hähnle, R., Huisman, M., Reger, G., Lidström, C.: Principles of Contract Languages (Dagstuhl Seminar 22451). Dagstuhl Reports, vol. 12, no. 11, pp. 1–27 (2023). https://doi.org/10.4230/DagRep.12.11.1

13. Hähnle, R., Huisman, M.: Deductive software verification: from pen-and-paper proofs to industrial tools. In: Steffen, B., Woeginger, G. (eds.) Computing and Software Science. LNCS, vol. 10000, pp. 345–373. Springer, Cham (2019). https://doi.org/10.1007/978-3-319-91908-9_18

14. Hoare, C.A.R.: The verifying compiler: a grand challenge for computing research. J. ACM **50**(1), 63–69 (2003). https://doi.org/10.1145/602382.602403

15. Huisman, M., Klebanov, V., Monahan, R.: On the organisation of program verification competitions. In: COMPARE. CEUR Workshop Proceedings, vol. 873, pp. 50–59 (2012). https://ceur-ws.org/Vol-873/papers/paper_2.pdf

16. Huisman, M., Monti, R., Ulbrich, M., Weigl, A.: The VerifyThis collaborative long term challenge. In: Ahrendt, W., Beckert, B., Bubel, R., Hähnle, R., Ulbrich, M. (eds.) Deductive Software Verification: Future Perspectives. LNCS, vol. 12345, pp. 246–260. Springer, Cham (2020). https://doi.org/10.1007/978-3-030-64354-6_10

17. Joshi, R., Holzmann, G.J.: A mini challenge: build a verifiable filesystem. Formal Asp. Comput. **19**(2), 269–272 (2007). https://doi.org/10.1007/s00165-006-0022-3

18. Klebanov, V., et al.: The 1st verified software competition: experience report. In: Butler, M., Schulte, W. (eds.) FM 2011. LNCS, vol. 6664, pp. 154–168. Springer, Heidelberg (2011). https://doi.org/10.1007/978-3-642-21437-0_14

19. Lanzinger, F., Weigl, A., Ulbrich, M., Dietl, W.: Scalability and precision by combining expressive type systems and deductive verification. Proc. ACM Program. Lang. **5**(OOPSLA), 1–29 (2021). https://doi.org/10.1145/3485520

20. Leavens, G.T., et al.: JML reference manual (2008)

21. Huismann, M., Monti, R.E., Ulbrich, M., Weigl, A. (eds.): VerifyThis Long-term Challenge 2020: Proceedings of the Online-Event (2020). https://doi.org/10.5445/IR/1000119426

22. Oortwijn, W., Gurov, D., Huisman, M.: Practical abstractions for automated verification of shared-memory concurrency. In: Beyer, D., Zufferey, D. (eds.) VMCAI 2020. LNCS, vol. 11990, pp. 401–425. Springer, Cham (2020). https://doi.org/10.1007/978-3-030-39322-9_19

23. Sprenger, C., et al.: Igloo: soundly linking compositional refinement and separation logic for distributed system verification. Proc. ACM Program. Lang. **4**(OOPSLA), 152:1–152:31 (2020). https://doi.org/10.1145/3428220

24. Stepney, S., Cooper, D., Woodcock, J.: An Electronic Purse: Specification, Refinement and Proof. Technical report PRG-126, Oxford University Computing Laboratory (2000). https://kar.kent.ac.uk/22009/1/An_Electronic_Purse_Specification,_Refinement_and_Proof.pdf

Author Index

D. Beyer et al. (Eds.): TOOLympics 2024, LNCS 14550, pp. 171–172, 2025.
https://doi.org/10.1007/978-3-031-67695-6